Robert Ferguson

The River-names of Europe

Robert Ferguson

The River-names of Europe

ISBN/EAN: 9783744791106

Printed in Europe, USA, Canada, Australia, Japan

Cover: Foto ©berggeist007 / pixelio.de

More available books at **www.hansebooks.com**

THE

RIVER-NAMES

OF

EUROPE.

BY ROBERT FERGUSON.

WILLIAMS & NORGATE,

14, HENRIETTA STREET, COVENT GARDEN, LONDON;

AND 20, SOUTH FREDERICK STREET, EDINBURGH,

CARLISLE: R. & J. STEEL.

1862.

PREFACE.

The object of the present work is to arrange and explain the names of European Rivers on a more comprehensive principle than has hitherto been attempted in England, or, to the best of my belief, in Germany.

I am conscious that, like every other work of the same sort, it must necessarily, and without thereby impugning its general system, be subject to correction in many points of detail. And in particular, that some of its opinions might be modified or altered by a more exact knowledge of the characteristics of the various rivers than can possibly in all cases come within the scope of individual research.

Among the writers to whom I am most indebted is Ernst Förstemann, who, in the second volume of his Altdeutsches Namenbuch, (the first consisting of the names of persons), has collected, explained, and where possible, identified, the ancient names of places in Germany. The dates affixed to most of the German rivers are taken from this work, and refer to the earliest mention of the name in charters or elsewhere.

I also refer here, because I find that I have not, as usual, given the titles elsewhere, to Mr. R. S. Charnock's "Local Etymology," and to the work of Gluck, entitled "Die bei C. Julius Cæsar vorkommende Keltische namen."

ROBERT FERGUSON.

CHAPTER I.

INTRODUCTION.

The first wave of Asian immigration that swept over Europe gave names to the great features of nature, such as the rivers, long before the wandering tribes that composed it settled down into fixed habitations, and gave names to their dwellings and their lands. The names thus given at the outset may be taken therefore to contain some of the most ancient forms of the Indo-European speech. And once given, they have in many, if not in most cases remained to the present day, for nothing affords such strong resistance to change as the name of a river. The smaller streams, variously called in England and Scotland brooks, becks, or burns, whose course extended but for a few miles, and whose shores were portioned out among but a few settlers, readily yielded up their ancient names at the bidding of their new masters.   But the river that flowed past, coming they knew not whence, and going they knew not whither—upon whose shores might be hundreds of settlers as well as themselves, and all as much entitled to give it a name as they—was naturally, as a matter of common convenience, allowed to retain its original appellation.

Nevertheless, it might happen that a river, such as the Danube, which runs more than a thousand miles as the crow flies—being divided between two great and perfectly distinct races, might, as it passed through the two different countries, be called by two different names.   So we find that while in its upper part it was called the Danube, in its lower part it was known as the Ister— the former, says Zeuss *(Die Deutschen)*, being its Celtic, and the latter its Thracian name.   So the Saone also was anciently known both as the Arar and the Sauconna —the latter, according to Zeuss, being its

Celtic name. And Latham, *(Tacitus, Germania,)* makes a similar suggestion respecting the Rhine—"It is not likely that the Batavians of Holland, and the Helvetians of Switzerland, gave the same name to the very different parts of their common river." It does not follow then as a matter of course—though we must accept it as the general rule—that the name by which a river is known at the present day, when it happens to be different from that recorded in history, is in all cases the less ancient of the two. There might originally have been two names, one of which has been preserved in history, and the other retained in modern use.

It is also to be observed, that in the case of one race coming after another—say Germans or Slaves after Celts—while the newcomers retained the old names, they yet often added a word of their own signifying water or river. The result is that many names are compounded of two words of different languages, and in not a few cases both signifying water.

The names thus given at the outset were
of the utmost simplicity, rarely, if ever, containing a compound idea. They were indeed
for the most part simple appellatives, being
most commonly nothing more than words
signifying water. But these words, once
established as names, entered into a different
category. The words might perish, but the
names endured. The words might change,
but the names did not follow their changes.
Inasmuch as they were both subject to the
same influences, they would most probably in
the main be similarly affected by them. But
inasmuch as the names were independent of
the language, they would not be regulated
in their changes by it. Moreover, in their
case a fresh element came into operation, for,
being frequently adopted by races speaking
a different language, they became subject to
the special phonetic tendencies of the new
tongue. The result is that many names,
which probably contained originally the
same word, appear in a variety of different

forms. The most important phonetic modifications I take to be those of the kind referred to in the next chapter.

There is no branch of philological enquiry which demands a wider range than that of the origin of the names of rivers. All trace of a name may be lost in the language in which it was given—we may have to seek for its likeness through the whole Indo-European family—and perhaps not find it till we come at last to the parent Sanscrit. Thus the name of the Humber is probably of Celtic origin, but the only cognate words that we find are the Latin *imber* and the Gr. ὄμβρος, till we come to the Sansc. *ambu*, water. Celtic also probably are the names of the Hodder and the Otter, but the words most nearly cognate are the Gr. ὕδωρ and the Lith. *audra*, (fluctus), till we come to the Sansc. *ud*, water.

Again, there are others on which we can find nothing whatever to throw light till we come to the Sanscrit. Such are the

Drave and the Trave, for which Bopp proposes Sansc. *dravas*, flowing. And the Arve
in Savoy, which I cannot explain till I come
to the Sansc. *arb* or *arv*, to ravage or destroy, cognate with Lat. *orbo*, Eng. *orphan*,
&c.   And—far as we have to seek for it—
how true the word is, when found, to the
character of that devastating stream ; and
how it will come home to the frequenters of
the vale of Chamouni, who well remember
how, within the last few years, its pretty
home-steads were rendered desolate, and
their ruined tenants driven out like "or-
phans" into the world! With such fury does
this stream, when swollen by the melted
snows, cast its waters into the Rhone, that
it seems to drive back the latter river into
the lake from whence it issues.   And Bullet
relates that on one occasion in 1572, the
mills of Geneva driven by the current of the
Rhone were made for some hours to revolve
in the opposite direction, and to grind their
corn backwards.

Thus then, though we may take it that
the prevailing element in the river-names
of Europe is the Celtic, we must turn for
assistance to all the languages that are cognate. And, for the double reason of their
great antiquity and their great simplicity,
we shall often find that the nearer we come
to the fountain-head, the clearer and the
more distinct will be the derivation. It will
be seen also throughout the whole of these
pages that, in examining the names of rivers,
we must take not only a wide range of philological enquiry, but also an extensive comparison of these names one with another.

The first step in the investigation is of
course to ascertain, whenever it is possible,
the most ancient forms in which these names
are found. We should scarcely suspect a
relationship between our Itchen and the
French Ionne, if we did not know that the
ancient name of the one was Icene, and of
the other Icauna. Nor would we suppose
that the Rodden of Shropshire was identical

with the French Rhone, did we not know that the original name of the latter was the Rhodănus.

In this, as in most other departments of philology, the industry of the Germans has been the most conspicuous. And Ernst Förstemann in particular, who has extracted and collated the ancient names of places in Germany up to the 12th cent., has furnished a store of the most valuable materials.

And yet after all there will be occasions on which all the resources of philology will be unavailing. Then we can but gather together the members of the family and wait till science shall reveal us something of their parentage. Thus the Alme that wanders among the pleasant meads of Devon—the Alm that flows by the quaint dwellings of the thrifty Dutch—the Alma that courses through the dark pine forests of the far North—the Almo that waters the sacred vale of Egeria—and the Alma, whose name brings sorrow and pride to many an English household—all contain one wide-spread and forgotten word, at the meaning of which we can but darkly guess.

CHAPTER II.

ON THE ENDINGS *a, en, er, es, et, el.*

We find that while there are many names of rivers which contain nothing more than the simple root from which they are derived, as the Cam, the Rhine, the Elbe, the Don, &c., there are others which contain the same root with various endings, of which the principal are *a, en, er, es, et, el.* Thus the Roth in Germany, contains a simple root ; the Roth(a), Roth(er), and Rodd(en) in England, and the Röt(el) in Germany, contain the same with four different endings. The German Ise shows a simple root, and the Germ. Is(ar), Is(en), Eng. Is(is), Dutch Yss(el), Russ. Iss(et), shew the same with five different endings. So we have in England the Tame, the Tam(ar), and the Tham(es), &c. The question is—what is the value and meaning of these various additions ?

With respect to the ending in *a*, found in some English rivers, there is reason to think that it is a word signifying water—the Old Norse *á*, Goth. *ahva*, Lat. *aqua*, &c. So that the *a* in Rotha may be the same as the *a* in the Norwegian Beina and the Swedish Tornea —as the *au* in the Germ. Donau (Danube) —and as the *ava* in the Moldava of Austrian Poland.

Others of these endings have by different writers been supposed to be also words signifying water. Thus Donaldson *(Varronianus)*, takes the ending *es* to have that meaning. And Förstemann, though more cautiously, makes the same suggestion for the termination *ar* or *er*. " I allow myself here the enquiry whether possibly the rivernames, which contain an *ar* as the concluding part of the word may not be compounded with this unknown word for a river ; to assume a simple suffix seems to me in this case rather niggardly." So also the ending *en* has been supposed by some of our own

Celtic scholars, as Armstrong and O'Brien, to be the same as the Welsh *aven*, Gael. *amhainn*, water or river, an opinion which has also, though to a more limited extent, received the sanction of Pott.

There are various minor objections to the above theories which I forbear to urge, because I think that the main argument against them is to be found in the manner in which these endings run through the whole European system of river-names. And it seems to me therefore more reasonable to refer them to a general principle which pervades the Indo-European languages, than to a particular word of a particular language. The principle I refer to is that of phonetic accretion, and it is that upon which the above word *aven* or *amhainn*, is itself formed from a simple root, by one of the very endings in question, that in *en*. Instead then of explaining—as the followers of the above system have done—the Saone (Sagonna), by the Celt. *sogh-an*, " sluggish river", I prefer to point to the general principle upon which the root *sogh* has the power, so to speak, of making itself into *soghan* (*e.g.* in Lat. *segn-is*.)

Not but that the principle contended for by the above writers may obtain in some cases : the Garumna, ancient name of the Garonne, looks like one of them, though even in this case I think that the latter may be the proper form, and the former only a euphonism of the Latin poets : the geographers, as Ptolemy, call it Garunna.

Then again the question arises whether, seeing that *en* and *es* in the Celtic tongues, and *el* in the Germanic, have the force of diminution, this may not be the meaning in the names of rivers. Zeuss, (*Die Deutschen*), suggests this in the case of the Havel and the Moselle ; but seeing that one of these rivers has a course of 180 and the other of 265 miles, I think they might rather be adduced to prove that these endings are not diminutive. We may cite also the Yssel and the Albula (Tiber), both large rivers, with this ending. While in Germany we have two rivers close together, the great and little Arl, (anc. Arla, or Arila)—here seems the very case for a diminutive, yet both rivers have the same ending. Not but that there are instances of a diminutive in river-names, but they seem of later formation. Thus there is no reason to doubt that the French Loiret, which is a small river falling into the large one, means "the little Loire." Etymology in this case is in perfect accord with the facts.

Upon the whole, then, I am inclined to the opinion, which seems in the main that of Förstemann, that, at least as the general rule, these endings are simply phonetic, and that they have no meaning whatever. In our own and the cognate languages, *en* is the principal phonetic particle—*e.g.*, English bow, Germ. bog*en*—Germ. rabe, Eng. rav*en* —Lat. virgo, Fr. vierge, Eng. virg*in*. But we have also traces in English of a similar phonetic *er*, *(see Latham's Handbook of* *the Eng. Language, p.* 199*).* The general reader will understand better what is here intended by comparing our words maid and maid*en.* Between these two words there is not the slightest shade of difference as regards meaning—the ending *en* is merely added for the sake of the sound, or, in other words, it is phonetic. Just the same difference then that there is between our words maid and maiden I take to be between the names of our rivers Lid and Lidden. The ending in both cases serves, if I may use the expression, to give a sort of finish to the word.

The question then arises—supposing these endings to be phonetic—were they given in the first instance, or have they accrued in after times? It is probable that both ways might obtain; indeed we have some evidence to shew that the latter has sometimes been the case. Thus the Medina in the Isle of Wight was once called the Mede, and the Shannon of Ireland stands in Ptolemy as

the Senus.   On the other hand cases are
more frequent in which the ending has been
dropped.   Thus the Yare is called by Ptolemy the Garrhuenus, *i.e.*, the Garron or
Yarron.   And the Teme appears in Anglo-
Saxon charters as the Taméde or Teméde.
Indeed the Thames itself would almost seem,
by having become a monosyllable, to have
taken the first step of a change which has
been arrested for ever.   So in Germany the
Bille, Ohm, Orre, and Bordau, appear in
charters of the 8th and 9th cent., as the
Bilena, Amana, Oorana, and Bordine.   And
in France the Isara and the Oscara have in
modern times become respectively the Oise
and the Ousche ; in both these two cases
the ending *er* has been dropped ; for Oise = *is*,
not *isar ;* and Ousche = *osc*, not *oscar*.

This latter principle is indeed only in
accordance with the general tendency of
language towards what Max Müller terms
"phonetic decay"—a principle which seems
less active in the rude than in the cultivated

stages of society. It would appear as if civilization sought to compensate itself for the increased requirements of its expression, by the simplification of its forms, and the rejection of its superfluous sounds.

Upon the whole then I think that as the general rule these endings have been given in the first instance, and that they have but rarely accrued in after times. Such being the case, though in one point of view they may be called phonetic, as adding nothing to the sense, yet in another point of view they may be called formative, as being the particles by means of which words are constructed out of simple roots. And of the names in the following pages, a great part, in some language, or in some dialect, are still living words. And those that are not, are formed regularly upon the same principle, common to the Indo-European system.

CHAPTER III.

ON THE MEANING OF RIVER-NAMES.

The names of rivers may be divided into two classes, appellative and descriptive—or in other words, into those which describe a river simply as "the water" or "the river," and those which refer to some special quality or property of its own.

In the case of a descriptive name we may be sure that it has been given—not from any fine-drawn attribute, but from some obvious characteristic—not from anything which we have to seek, but from something which, as the French say, "saute aux yeux." If a stream be very rapid and impetuous— if its course be winding and tortuous—if its waters be very clear or very turbid— these are all marked features which would naturally give it a name.

But such derivations as the following from Bullet can only serve to provoke a smile. Thus of the Wandle in Surrey he says—" Abounding in excellent trouts—*van*, good, *dluz*, a trout." (I much fear that the "excellent trouts" have been made for the derivation, and not the derivation for the trouts.) Of the Irt in Cumberland he says —" Pearls are found in this river. Irt signifies surprising, prodigious, marvellous." Marvellous indeed ! But Bullet, though nothing can be more childish than many of his etymological processes, has the merit of at least taking pains to find out what is actually the notable feature in each case under consideration, a point which the scholarly Germans sometimes rather neglect.

River-names, in relation to their meaning, may be ranked under seven heads.

1. Those which describe a river simply as " the water," " the river." Parallel with this, and under the same head, we may take the words which describe a river

as " that which flows," because the root-meaning of most of the words signifying water is, that which flows, that which runs, that which goes. Nevertheless, there may be sometimes fine shades of difference which we cannot now perceive, and which would remove the names out of this class into the next one.

2. Those which, passing out of the appellative into the descriptive, characterize a river as that which runs violently, that which flows gently, or that which spreads widely.

3. Those which describe a river by the nature of its course, as winding, crooked, or otherwise.

4. Those which refer to the quality of its waters, as clear, bright, turbid, or otherwise.

5. Those which refer to the sound made by its waters.

6. Those which refer to the nature of its

source, or the manner of its formation,
as by the confluence of two or more
streams.

7. Those which refer to it as a boundary or
as a protection.

Under one or other of the above heads
may be classed the greater part of the river-
names of Europe.

And how dry and unimaginative a list it
is ! We dive deep into the ancient language
of Hindostan for the meaning of words, but
we recall none of the religious veneration to
the personified river which is so strikingly
manifest even to the present day. As we
read in the Vedas of three thousand years
ago of the way-farers supplicating the spirit
of the stream for a safe passage, so we read
in the newspapers of to-day of the pilgrims,
as the train rattled over the iron bridge,
casting their propitiatory offerings into the
river below. We seek for word-meanings in
the classical tongue of Greece, but they
come up tinged with no colour of its grace-

ful myths. Few and far between are the
cases—and even these are doubtful, to say
the least—in which anything of fancy, of
poetry, or of mythology, is to be traced in
the river-names of Europe.

CHAPTER IV.

APPELLATIVES.

The great river of India, which has given its name to that country, is derived from Sansc. *sindu*, Persian *hindu*, water or sea. It was known to the ancients under its present name 500 years B.C. Another river of Hindostan, the Sinde, shews more exactly the Sansc. form, as the Indus does the Persian. It will be seen that there are some other instances of this word in the ancient or modern river-names of Europe.

1. *India.*   The INDUS and the SINDE.
   Asia Minor.   INDUS ant., now the Tavas.
   France.   INDIS ant., now the Dain.
   Germany.   INDA, 9th cent. The INDE near Aix-la-Chapelle.
   Norway.   The INDA.
2.   *With the ending er.*
   France.   The INDRE. Joins the Loire.

The most widely spread root is the Sansc. *ap*, Goth. *ahva*, Old High Germ. *aha*, Old

Norse *á*, Ang.-Sax. *ea*, Lat. *aqua*, &c. With
the form *ahva* Fürst connects Ahava as the
name of a river in the district of Babylon,
mentioned in Ezra, chap. 8, v. 21—"Then
I proclaimed a fast there at the river of
Ahava." But from the 15th verse it would
rather seem that Ahava was a place and not
a river—"and I gathered them together to
the river that *runneth to* Ahava." The place
might certainly, as in many other cases, take
its name from the river on which it stood,
but this is one step further into the dark.
From the root *ab* or *ap* is formed Latin
amnis, a river, corresponding, as Diefenbach
suggests, with a Sansc. *abnas*. Also the
Celt. *auwon, avon, abhain*, or *amhain*, of the
same meaning, from the simple form found
in Obs. Gael. *abh*, water. The Old German
aha, awa, ava, or *afa*, signifying water
or river, is added to many names of that
country which are themselves probably of
Celtic or other origin ; the form in Modern
German is generally *ach* or *au*. The ending

in *a* of some English rivers, as the Rotha, Bratha, &c., I have already suggested, chapter 3, may be from the same origin; this form corresponds most nearly with the Scandinavian. There are one or two, as the Caldew in Cumberland, which seem to show the Germ. form *au* or *ow*. The ending *ick* or *ock* in several Scotch rivers, as the Bannock and the Errick, may be from a word of similar meaning, most probably the obs. Gael. *oich*.

I divide the widely spread forms from this root for convenience into two groups, *ap* or *av*, and *ach* or *ah*. The relation between the consonants is shown in the Gr. ῖππος, Lat. *equus*, Ang.-Sax. *eoh*, horse, three words similarly formed from one root. The European names in the following group I take to be most probably from the Celtic—the Asiatic, if they come in, must be referred to the Sanscrit, or a kindred and coeval tongue.

1. *England.* The IVE. Cumberland.
   Portugal, The AVIA.

Germany.	IPFA, 8th cent., now the IPF—here?
Asia Minor.	HYPIUS ant.—here?

2. *With the ending en = Celtic auwon, avon, abhain, amhain, Lat. amnis.*

England.	The AVON and EVAN. Many rivers in England, Scotland, and Wales.
Scotland.	The AMON, near Edinburgh, also, but less correctly, called the ALMOND.
France.	The AVEN., Dep. Finistère.
Germany.	AMANA, 8th cent., now the OHM.
Hindostan.	HYPANIS ant., now the Sutledge—here?
Asia Minor.	EVENUS ant., now the Sandarli—here? AMNIAS ant., probably here.
Syria.	ABANA ant., now the Barrada—here?

3. *With the ending er.*

France.	The AVRE. Dep. Eure.
Germany.	IVARUS, 2nd cent., now the Salzach. EPAR(AHA), 8th cent., now the EBR(ACH).
Spain.	IBERUS ant., now the EBRO.
Thrace.	HEBRUS ant., now the Maritza.

4. *With the ending el.*

England.	The IVEL.* Somers.
Germany.	APULA, 9th cent. The APPEL(BACH)

* Ilchester (=Ivel-chester) situated on this river, is called in Ptolemy Ischalis, from which we may presume that the river was called the Ischal, a word which would be a synonyme of Ivel.

Hungary.	The IPOLY or EYPEL. Joins the Danube.

5.   *With the ending es.* *

Germany.	IBISA, 8th cent. The IPS.
Portugal.	The AVIZ.
Sicily.	HYPSAS ant., now the Belici.
Illyria.	APSUS ant., now the Beratinos.

A related form to No. 2 of the above group I take to be *ain* = Manx *aon* for *avon.*

England.	The AUNE, Devonshire. The EHEN, Cumberland. The INNEY, Cornwall.
Germany.	The AENUS of Tacitus, now the INN. The IHNA, Prussia.
Greece.	OENUS ant.—here ?

And I place here also a form *annas,* which I take to be = Sansc. *abnas,* Latin *amnis.*

India.	The ANNAS. Gwalior.
Germany.	ANISA, 8th cent. The ENS in Austria.
Piedmont.	The ANZA. Joins the Tosa.

In the other form *ah, ach,* there may be more admixture of the German element.

---

* It seems rather probable that the ending *es* in these names is not a mere suffix. The APSARUS, ancient name of the Tchoruk in Armenia, and the IPSALA in Europ. Turkey, by superadding the endings *er* and *el,* go to show this. We might perhaps presume a Sansc. word *abhas* or *ophas,* with the meaning of river.

But the English names, I take it, are all
Celtic. The form *ock* comes nearest to the
obs. Gael. *oich.*

1. *England.* The OCK, Berks. The OKE, Devon.
   Scotland. The OICH, river and lake. The
           AWE, Argyle. The EYE, Berwicks.
   France. The AA. Dep. Nord.
   Germany. The AACH and the AU.
   Holland. The AA in Brabant.
   Russia. The OKA and the AA.
2.   *With the ending el.*
   Scotland. The OIKELL. Sutherland.
   Germany. AQUILA, 8th cent., now the EICHEL.

With the Sanscrit root *ab* or *ap* is to be
connected Sanscrit *ambu, ambhas,* water,
whence Latin *imber* and Gr. ὄμβρος. If the
Abus of Ptolemy was the name of the river
Humber, it contains the oldest and simplest
form of the root. But the river is called
the Humbre in the earliest Ang.-Sax. records.
I class in this group also the forms in *am*
and *em.*

1. *England.* The EMME. Berkshire.
   Switzerland. The EMME.
   Holland. EMA, 10th ct., now the EEM—here ?
   Sweden. The UMEA.

Asia. The EMBA, also called the Djem.

2. *With the ending en.*
Switzerland. The EMMEN. Two rivers.

3. *With the ending er.*
England. The HUMBER. Humbre, *Cod. Dip.*
The AMBER. Derbyshire.
Germany. AMBRA, 8th cent., now the AMMER, and the EMMER.
Italy. UMBRO ant., now the OMBRONE.

4. *With the ending el.*
England. The AMBLE or HAMBLE. Hants.
The AMELE or EMELE, now the Mole, in Surrey.
Germany. The HAMEL. Hanover.
Belgium. AMBL(AVA), 9th cent., now the AMBL(EVE).

5. *With the ending es, perhaps = Sansc. ambhas, water.*
England. The HAMPS. Stafford.
France. The AMASSE. Joins the Loire.
Germany. AMISIA, 1st cent. The EMS in Westphalia.
EMISA, 8th cent. The EMS in Nassau.

6. *With the ending st.**
Asia. AMBASTUS ant. Now the Camboja.

---

* This ending is not explained. Zeuss, comparing the endings *er* and *st*, suggests a comparative and superlative, which is not probable. In the present, as in some other cases, I take it to be only a phonetic form of *ss*, and make Ambastus properly Ambassus. But in some other cases, as that of the Nestus, which compares with Sansc. *nisitas*, fluid, it seems to be formative.

The whole of the above forms are to be traced back to the Sanscrit verb *ab* or *amb*, signifying to move ; and that probably to a more simple verb *á*. The Old Norse *á*, Ang.-Sax. *eá*, water or river, contain then a root as primitive as language can show. We can resolve it into nothing simpler—we can trace it back to nothing older. And it is curious to note how the Latin *aqua* has, in the present French word *eau*, come round again once more to its primitive simplicity. Curious also to note to what phonetic proportions many of the words, as the Avon, the Humber, &c., have grown, and yet without adding one particle of meaning, as I hold, to the primeval *á*.

The root of the following group seems to be Sansc. *ux* or *uks*, to water, whence Welsh *wysg*, Irish *uisg*, Old Belg. *achaz*, water or river. Hence also Eng. *ooze*, and according to Eichoff (*Parrallele des langues*), also *wash*.

1. *England.* The AXE, Devon. The AXE, Somers.

England.	The Ash, Wilts. *Cod. Dip.* Asce.
	The Isaca, or Isca (Ptolemy). The Exe.
	The Esk, Cumb. Eske, Yorks.
	The Esk, in Scotland, five rivers.
	The Usk, in Monmouthshire.
France.	The Isac. Dep. Mayenne.
	The Esque. Normandy.
	The Achase. Dauphiné.
Germany.	Achaza, 10th cent., now the Eschaz.
	Acarse,† 11th cent., now the Axe.
	The Ahse. Prussia.
Mœsia.	Œscus ant.
Asia.	Aces ant. (Herodotus), now the Oxus or Amou.
Greece.	Axius ant., now the Vardar in Macedon.* Axus or Oaxes in Crete, still retains its name.

2.		*With the ending en.*
	France.	Axona ant. (Cæsar.) Now the Aisne.
	Asia.	Ascania ant. Two lakes, one in Phrygia, and the other in Bithynia.

3.		*With the ending el.*
	England.	Uxella ant., (Richard of Cirencester), supposed to be the Parret.
		The Eskle, Hereford.
	Germany.	Iscala, 8th cent. The Ischl.
	Russia.	The Oskol. Joins the Donetz.

† This looks like a mistake for Acasse.

* So that there *is* a river in Monmouth, and another in Macedon.

4.   *With the ending er.*
France.   OSCARA ant., now the OUSCHE.
Belgium.   HISSCAR, 9th cent., seems not to be
              identified.

I am inclined to bring in here the root *is*,
respecting which Förstemann observes that
it is "a word found in river-names over a
great part of Europe, but the etymology of
which is as yet entirely unknown." I connect it with the above group, referring also
to the Old Norse *is* motus, *isia*, proruere, as
perhaps allied. I feel an uncertainty about
bringing the name OUSE either in this group
or the last, for two at least of the rivers so
called are so very tortuous in their course
as to make us think of the Welsh *osgo*,
obliquity.

1. *Germany.*   The ISE and the EIS(ACH).
   Syria.   ISSUS ant., now the Baias—here?

2.   *With the ending en.*
   Germany.   ISANA, 8th cent.   The ISEN.

3.   *With the ending er.*
   France.   ISARA, 1st cent. B.C.   The ISÈRE
                and the OISE.*

---

* "Hysa nunc fluvii !nomen est, qui antiquitus Hysara dicebatur."
(*Folcuin. Gest. Abb. Lobiens.*) This seems not improbably to refer to the
Oise.

| *Germany.* | ISARA ant.   The ISAR. |

4. *With the ending el.*
Scotland. The ISLA.   Two rivers.
France. The ISOLÉ.
Holland. ISELA, 8th cent., now the YSSEL.
Spain. The ESLA.

5. *With the ending es.*
England. The ISIS, vulg. Ouse.

6. *With the ending et.*
Siberia. The ISSET.   Joins the Tobol.

7. *In a compound form.*

The ISTER, or Danube, perhaps = IS-STER, from a word *ster*, a river, hereafter noticed.

ISMENUS ant., in Bæotia.   The ending seems to be from a Celt. word *man* or *mon*, probably signifying water or river, and found in several other names, as the Idumania of Ptolemy, now the Blackwater, the Alcmona of Germany, now the Altmühl, the Haliacmon of Macedonia, now the Vistritza, &c.

HESUDROS, the ancient name of the Sutledge (Sansc. *udra*, water), may also come in.

From the Sansc. *ud*, water—in comp. *udra*, as in *samudra*, the sea, *i.e.*, collection of waters, (see also Hesudros above)—come Sansc. *udon*, Gr. ʽύδωρ, Slav. *woda*, Goth. *wato*, Germ. *wasser*, Eng. *water*, Lith. *audra*, fluctus, &c.

E

1. *Italy.*   ADUA ant., now the ADDA.
   Bohemia.   The WAT(AWA).

2.   *With the ending en = Sansc. udon, water ?*
   France.   The ODON.
   Germany.   ADEN(OUA), 10th cent., now the
                 ADEN(AU).

3. *With the ending er = Germ. wasser, Eng. water,*
                 &c.
   England.   The ODDER and the OTTER.
                 The WODER, Dorset. Woder, *Cod.
                 Dip.*
                 The ADUR in Sussex.
                 The VEDRA of Ptolemy, now the Wear,
                 according to Pott, comes in here.
   France.   ATURUS ant., now the ADOUR.
                 AUDURA ant., now the EURE.
   Germany.   ODORA ant., now the ODER.
                 WETTER(AHA), 8th cent., now the
                 WETTER.*

4.   *With the ending rn.†*
   Germany.   ADRANA, 1st cent., now the EDER.
   Asia Minor. The EDRENOS.   Anc. Rhyndacus.

5.   *With the ending el.*
   Russia.   The VODLA.   Lake and river.

---

* If, as Pott suggests, the Vedra of Ptolemy=Eng. *water*, the Wetter
would naturally come in here also. But some German writers, as Roth
and Weigand, connect it with Germ. *wetter*, Eng. *weather*, in the sense,
according to the first-named, of the river which is affected by rain.

† This ending may either be formed by the addition of a phonetic *n* to
the ending *er*; or it may be from a word *ren*, channel, river, hereafter
noticed.

To the above root I also put a form in *ed*, corresponding with Welsh *eddain*, to flow, Ang.-Sax. *edre*, a water-course, &c.

1.   *With the ending en.*
England. The EDEN. Cumberland. Probably the Ituna of Ptolemy.
Scotland. The EDEN and the YTHAN.
France. The ITON. Joins the Eure.

2.   *With the ending er.*
Scotland. The ETTR(ICK). Joins the Tweed.
Germany. EITER(AHA), 8th cent. The EITR(ACH)* the EITER(ACH), and the AITER-(ACH).
Denmark. EIDORA ant., now the EIDER.

3.   *With the ending el.*
England. The IDLE. Notts.

4.   *With the ending es.*
Germany. IDASA, 11th cent., now the ITZ.

With the above may perhaps also be classed the Celtic *and* or *ant*,† to which Mone, (*Die Gallische sprache*), gives the meaning of water.

1. *England.* The ANT. Norfolk.

---

* The Scotch EITRICK and the Germ. EITRACH I take to be synonimous, though the ending in one case is German, and in the other probably Gaelic. (*See p.* 25)

† Hence perhaps Anitabha (*abha*, water), the Sansc. name of a river, not identified, in India.

2.		*With the ending en.*
	England.	The Anton.* Hants.
3.		*With the ending er.*
	France.	Andria ant. Now the Lindre.
4.		*With the ending el.*
	France.	The Andelle. Joins the Seine.
	Germany.	Antil(aha), 10th cent., now the Andel(au).

To the Celt. ˈdubr, Welsh *dwfr*, water, are by common consent referred the names in the second division of the undermentioned. But the forms *dub, duv,* which in accordance with the general system here advocated, I take to be the older and simpler form of the word, are, by Zeuss (*Gramm. Celt.*), as well as most English writers, referred to Welsh *du,* Gael. *dubh,* black.

1. *England.*	The Dove. Staffordshire.
	The Dow. Yorkshire.
Wales.	Tobius ant., now the Towy.
	The Dovy, Merioneth.
France.	Dubis ant., now the Doubs.
	The Doux, joins the Rhine.

---

* Tacitus gives this name to the Avon—in mistake, as the Editor of Smith's Ancient Geography suggests. But *anton* and *avon* seem to have been synonimous words for a river.

2. *With the ending er, forming the Celtic dubr,*
*Welsh dwfr.**

Ireland.	Dobur ant., retains its name.†
France.	The Touvre.
Germany.	Dubra, 8th cent., now the Tauber.
	The Daubr(awa), Bohemia.
3.	*With the ending es.*
Russia.	The Dubissa.

Another Celtic word for water is *dur*, which, however, seems more common in the names of towns (situated upon waters), than in the names of rivers. Is this word formed by syncope from the last, as *duber* = *dur*? Or is it directly from the root of the Sansc. *drâ* or *dur*, to move?

1. *England.*	The Durra.	Cornwall.
Germany.	Δοῦρας, Strabo, now the Iller or the Isar.	
Switz.	Dura, 9th cent. The Thur.‡	
Italy.	Duria ant., now the Dora.	
	Turrus ant., now the Torre.	
Spain.	Durius ant., now the Douro.	
Russia.	The Tura. Siberia.	
	The Turija. Russ. Poland.	

---

* Hence the name of Dover, anc. Dubris, according to Richard of Cirencester, from the small stream which there falls into the sea.

† Where is this river, cited by Zeuss, (*Gramm. Celt.*)?

‡ Hence probably the name of Zurich, ant. Turicum.

2. *With the ending en.*

France. DURANIUS ant., now the DORDOGNE.

In this chapter is to be included the root *ar*, respecting which I quote the following remarks of Förstemann. "The meaning of river, water, must have belonged to this wide-spread root, though I never find it applied as an appellative, apart from the obsolete Dutch word *aar*, which Pott produces. I also nowhere find even an attempt to explain the following river-names from any root, and know so little as scarcely to make a passing suggestion; even the Sanscrit itself shows me no likely word approaching it, unless perhaps we think of *ara*, swift (*Petersburger Wörterbuch*)."

The root, I apprehend, like that of most other river-names, is to be found in a verb signifying to move, to go—the Sansc. *ar*, *ir* or *ur*, Lat. *ire*, *errare*, &c. And we are not without an additional trace of the sense we want, as the Basque has *ur*, water, *errio*, a river, and the Hung. has *er*, a brook. The sense of swiftness, as found in Sansc. *ara*, may perhaps intermix in the following names. But there is also a word of precisely opposite meaning, the Gael. *ar*, slow, whence Armstrong, with considerable reason, derives the name of the Arar (or Saone), a river noted above all others for the slowness of its course. Respecting this word as a termination see page 11.

1. *England.*	The ARROW, Radnor. The ARROW, Worcester.
	The ORE. Joins the Alde.
Ireland.	ARROW, lake and river, Sligo.
France.	The AURAY. Dep. Morbihan.
Germany.	ARA, 8th cent. The AHR, near Bonn, the OHRE, which joins the Elbe, and the OHRE in Thuringia, had all the same ancient name of Ara.
	UR(AHA), 10th cent., now the AUR(ACH).
Switzerland.	ARA, ant. The AAR.
Italy.	The ERA. Joins the Arno.
Spain.	URIUS ant., now the Rio Tinto.
Russia.	OARUS (Herodotus), perhaps the Volga.
2.	*With the ending en.*
England.	The ARUN, Sussex.

Scotland.	The ORRIN and the EARNE.
Ireland.	The ERNE, Ulster.
Germany.	OORANA, 8th cent., now the ORRE.
	ARN(APE), 8th cent., (*ap,* water), now the ERFT.
	The OHRN. Wirtemberg.
Tuscany.	ARNUS ant. The ARNO.

3.   *With the ending el.*

Germany.	ERL(AHA), 11th cent. The ERLA.
	URULA, 9th cent. The ERL.
	ARLA, 10th cent. The ARL.
	The ORLA. Joins the Saale.
Savoy.	The ARLY.
Aust. Slavonia.	The ORLY(AVA).
Russia.	The URAL and the ORL(YK).

From *ár* and *ur,* to move, the Sanscrit forms *arch* and *urj,* with the same meaning, but perhaps in a rather more intense degree, if we may judge by some of the derivatives, as Lat. *urgeo,* &c. In two of the three appellatives which I find, the Basque *erreca,* brook, and the Lettish *urga,* torrent, we may trace this sense ; but in the third, Mordvinian (a Finnish dialect), *erke,* lake, it is altogether wanting. And on the whole, I cannot find it borne out in the rivers quoted below. Perhaps the Obs. Gael. *arg*, white, which has been generally adduced as the etymon of these names, may intermix.

1. *England.*	The ARKE.	Yorkshire.
	The IRK.	Lancashire.
France.	The OURCQ.	Dep. Aisne.
	The ORGE and the ARC.	
Belgium.	The HERK.	Prov. Limburg.
Sardinia.	The ARC.	Joins the Isere.
Spain.	The ARGA.	Joins the Aragon.
Armenia.	ARAGUS ant., now the ARAK.	
2.	*With the ending en.*	
Germany.	ARGUNA, 8th cent.	The ARGEN.
Russia.	The ARGUN.	Two rivers.
Spain.	The ARAGON.	Joins the Ebro.
3.	*With the ending et.*	
Siberia.	The IRKUT.	Joins the Angara.
4.	*With the ending es.*	
France.	The ARQUES.	
Russia.	The IRGHIZ.	Two rivers.
5.	*With the ending enz.*[*]	
Germany.	ARGENZA, 9th cent., now the ERGERS.	

From the Sansc. *ri*, to flow, Gr. ῥεω, Lat. *rigo* (often applied to rivers—" Qua Ister Getas rigat," *Tibullus*), Sansc. *rinas*, fluid, Old Sax. *riha*, a torrent, Ang.-Sax. *regen*,

---

* Perhaps formed from *ez* by a phonetic *n*.

F

Eng. *rain*, Slav. *rêka*, a stream, Welsh *rhe*, rapid, *rhean*, *rhen*, a stream, &c., we get the following group. The river Regen Berghaus derives from Germ. *regen*, rain, in reference to the unusual amount of rain-fall which occurs in the Böhmer-wald, where it has its source. Butmann derives it from Wend. and Slav. *rêka*, a stream, connecting its name also with that of the Rhine. Both these derivations I think rather too narrow.

With respect to the Rhine I quote the following opinions. Armstrong derives it from Celt. *reidh-an*, a smooth water, than which nothing can be more unsuitable—the characteristic of the river, as noticed by all observers, from Cæsar and Tacitus downwards—being that of rapidity. Donaldson compares it with Old Norse *renna*, fluere, and makes Rhine = Anglo-Saxon *rin*, cursus aquæ. Grimm (*Deutsch. Gramm.*) compares it with Goth. *hrains*, pure, clear, and thinks that "in any case we must dismiss the derivation from *rinnan*, fluere." Zeuss and Förstemann support the opinion of Grimm ; nevertheless, all three agree in thinking that the name is of Celtic origin. The nearest word, as it seems to me, is Welsh *rhean*, *rhen*, a stream, cognate with Sansc. *rinas*, fluid, Old Norse *renna*, fluere, and (as I suppose), with Goth. *hrains*, pure.

1. *England.*	The REA.	Worcester.
	The WREY.	Devonshire.
Ireland.	The RYE.	Joins the Liffey.
Germany.	The REGA.	Pomerania.
Holland.	The REGGE.	Joins the Vecht.
Spain.	The RIGA.	Pyrenees.
Russia.	RHA ant., now the Volga.	
2.	*With the ending en.*	
Germany.	REGIN, 9th cent. The REGEN.	
	RHENUS, 1st cent. B.C. The RHINE.	
	The RHIN. Joins the Havel.	
	The RHINE. A small stream near Cassel.	
Norway.	The REEN.	
Italy.	The RENO by Bologna.	
Asiat. Russ.	The RHION, ant. Phasis.	

The Sansc. *li*, to wet, moisten, spreads into many forms through the Indo-European languages. I divide them for convenience into two groups, and take first Lat. *liqueo*, Old

Norse *leka*, Ang.-Sax. *lecan* (stillare, rigare), Gael. and Ir. *li*, sea, Gael. *lia*, Welsh *lli*, *llion*, a stream. Most of the following names, I take it, are Celtic. I am not sure that the sense of stillness or clearness does not enter somewhat into the two following groups.

1. *England.*   The LEE.   Cheshire.
            The LEACH.   Gloucestershire.
Ireland.   The LEE.   Two rivers.
Germany.   LICUS, 2nd cént., now the LECH.
            LIA, 8th cent., now the LUHE.
France.   LEGIA, 10th cent., now the LYS.*
Belgium.   The LECK.   Joins the Maas.
Hindostan. The LYE.   Bengal.

2.   *With the ending en = Welsh llion, a stream.*
England.   The LEEN.   Notts.
Scotland.   The LYON and the LYNE.
France.   The LIGNE.   Dep. Ardéche.

3.   *With the ending er.*
England.   The LEGRE by Leicester, now the Soar.
France.   LIGER ant.   The LOIRE.
            The LEGRE.   Dep. Gironde.

For the second group I take Lat. *lavo, luo,* Old Norse *lauga*, lavare, Anglo-Saxon *lagu*,

---

* I do not in this case make any account of the spelling; the name is just the same as our Lee, and the idea of *lys*, a lily, is no doubt only suggested by the similarity of sound.

water, Gael. *lo*, water, Gael. and Ir. *loin*, stream. In this group there may perhaps be something more of the German element, *e. g.*, in the rivers of Scandinavia.

1.	*England.*	The LUG.	Hereford.
	Wales.	The LOOE.	Two rivers.
	France.	The LOUE.	Dep. Haute Vienne.
	Germany.	LOUCH(AHA), 11th cent. The LAUCHA.	
		LOUA, 10th cent., not identified.	
	Holland.	The LAVE.	
	Finland.	The LUGA or LOUGA.	
2.		*With the ending en.*	
	England.	The LUNE.	Lancashire.
		The LAINE.	Cornwall.
		The LEVEN.	Two rivers.
	Scotland.	The LEVEN.	Two rivers.
	Ireland.	The LAGAN, near Belfast.	
	France.	LUNA ant., now the LOING.	
	Germany.	LOGAN(AHA), 8th cent., now the LAHN.	
		The LOWNA in Prussia.	
	Norway.	The LOUGAN.	Joins the Glommen.
		The LOUVEN.	Stift Christiania.
	Russia.	The LUGAN.	
	Italy.	The LAVINO.	
		The lake LUGANO.	
	India.	The LOONY—here ?	
3.		*With the ending er.*	
	Scotland.	The LUGAR.	Ayr.
	Wales.	The LLOUGHOR.	Glamorgan.

To the above root I also place the following, corresponding more distinctly with Welsh *llifo*, to pour.

1. *Ireland.* The LIFFEY by Dublin.
   Germany. LUPPIA, 1st cent. The LIPPE.
   The LIP(KA). Bohemia.
2. *With the ending er.*
   England. The LIVER. Cornwall.
   Scotland. The LIVER. Argyle.
   Ireland. The LIFFAR.

More remotely with the Sansc. *li*, liquere, and directly with Welsh *lleithio*, to moisten, *llyddo*, to pour, Gael. *lith*, a pool, smooth water, Goth. *leithus*, Ang.-Sax. *lidh*, liquor, poculum, potus, I connect the following. The rivers themselves hardly seem to bear out the special idea of smoothness, which we might be apt to infer from the root, and from the character of the mythological river Lethe.

1. *England.* The LID. Joins the Tamar.
   Scotland. The LEITH. Co. Edinburgh.
   Wales. The LAITH, now called the Dyfr.
   Germany. LIT(AHA), 11th cent. The LEITHA.
   Sweden. The LIDA.
   Hungary. The LEITHA. Joins the Danube.

Asia Minor.
Thessaly. } LETHÆUS ant., three rivers—here ?
Crete.

2.   *With the ending en.*
England.   The LIDDEN (Leden, *Cod. Dip.*)
                     ˙Worcester.
Scotland.   The LEITHAN.   Peebles.
3.   *With the ending el.*
Scotland.   The LIDDLE.   Joins the Esk.

From the Sansc. *nî*, to move, comes *nîran*, water, corresponding with the Mod. Greek νερόν of the same meaning. And that the Greek word is no new importation into that language, we may judge by the name of Nereus, a water-god, the son of Neptune. The Gr. *ναω*, fluo, the Gael. *nigh*, to bathe, to wash, and the Obs. Gael. *near*, water, a river, show a close relationship ; the Heb. *nhar*, a river, also seems to be allied. Compare the Nore, a name given to part of the estuary of the Thames, with the Narra, the name of the two branches by which the Indus flows into the sea. Also with the Nharawan, an ancient canal from the Tigris towards the Persian Gulf. And with the

Curische Nehrung, a strip of land which separates the lagoon called the Curische Haf in Prussia from the waters of the Baltic. On this name Mr. Winning remarks,* "I offer the conjecture that the word *nehrung* is equivalent to our break-water, and that it is derived from the Sabine (or Old Prussian) term *neriene*, strength, bravery." I should propose to give it a meaning analagous, but rather different—deriving it from the word in question, *nar* or *ner*, water, and some equivalent of Old Norse *engia*, coarctare, making *nehrung* to signify "that which confines the waters" (of the lake). In all these cases there is something of the sense of an estuary, or of a channel communicating with the sea—the Curische Haf being a large lagoon which receives the river Niemen, and discharges it by an outlet into the Baltic. The following names I take to be for the most part of Celtic origin.

---

* Manual of Comparative Philology.

1. *England.* The Now. Derbyshire.

The NAR. Norfolk.

The NORE, part of the estuary of the Thames.

Ireland. NEAGH. A lake, Ulster.

NORE. Joins the Shannon.

Germany. NOR(AHA), 8th cent., also called the NAHA.

Italy. NAR,* ant. The NERA.

Spain. The NERJA. Malaga.

Russia. The NAR(OVA), and the NAREW.

Europ. Turkey. NARO ant., now the NARENTA.

Mauretania. NIA ant., now the Senegal—here?

Hindostan. NARRA, two branches of the Indus—here?

2. *With the ending en, = Sansc. nîran, water?*

Illyria. The NARON.

Scotland. The NAREN or NAIRN.

3. *With the ending es.*

Germany. The NEERS. Rhen. Pruss.

From the Sansc. *nî*, to move, Gael. *nigh*, to bathe, to wash, comes, I apprehend, the Welsh *nannaw, nennig, nant*, a small stream.

England. The NENE or NEN. Northampton.

The NENT. Cumberland.

Ireland. The NENAGH. Joins the Shannon.

France. The NENNY.

---

* Niebuhr derives this name from a Sabine word signifying sulphur, which is largely contained in its waters. Mr. Charnock suggests the Phœn. *naharo*, a river.

Closely allied to *ni*, to move, I take to be
Sansc. *niv*, to flow, Welsh *nofio*, to swim, to
float, whence the names undermentioned. The
Novius of Ptolemy, supposed to be the Nith,
if not a false rendering, might come in here.

1. *France.*   The NIVE. Joins the Adour.
    Germany.   NABA, 1st cent., now the NAAB in Bavaria.
    Holland.   NABA or NAVA, 1st cent., now the NAHE or NAVE.
    Spain.   The NAVIA. Falls into the Bay of Biscay.
    Russia.   The NEVA and the NEIVA.
    Hindostan.   The NAAF. Falls into the Bay of Bengal.

2.   *With the ending en.*
    Persia.   The NABON. Prov. Fars.
    Russ. Pol.   The NIEMEN.*

3.   *With the ending er.*
    Scotland.   The NAVER. River and lake.
    Wales.   The NEVER. Merioneth.
    France.   NIVERIS ant., now the NIEVRE.
    Danub. Prov. NAPARIS (Herodotus), supposed to be the Ardisch.

4.   *With the ending el.*
    France and Spain. } The NIVELLE. Pyrenees.

---

* Niemen may perhaps=Nieven—*m* for *v*, as in Amon for Avon, p. 26.

Holland. NABALIS (Tacitus), by some thought to be the Yssel.

5.   *With the ending es.*

Scotland. The NEVIS. Rises on Ben Nevis.

From the same root, *nĭ*, to move, and closely connected with the last group, I take to be Sansc. *nis*, to flow, to water. Zeuss (*Die Deutschen*) takes the word, as far as it relates to the rivers of Germany, to be of Slavonic origin. It appears to be the word found as the second part of some Slavonic river-names, as the Yalomnitza. But it is also both Celtic and Teùtonic, for the Armorican has *naoz*, a brook, and the German has *nasz*, wet, *nässen*, to be wet.

1. *Scotland.* The NESS. River and lake.

   Germany. NISA, 11th cent. The NEISSE, two rivers, both of which join the Oder.

   Servia. The NISS(AVA). Joins the Morava.

   Sicily. The NISI.

2   *With the ending st.**

   France. The NESTE. Hautes Pyrenees.

   Thrace. NESTUS ant.

---

* Perhaps to be found in Sansc. *nistas*, wet, fluid. Here we get something of a clue to Eng. "nasty," the original meaning of which has no doubt been nothing but water " in the wrong place. '

From the Greek *ναω*, fluo, comes *νᾶμα*, a stream, *ναματιᾶιον* ʿύδωρ, running water. Hence seems to be NAMADUS, the name given by the Greek geographers to the Nerbudda of India.

Another form which I take to be derived from the above Sanscrit root *nî*, by the prefix *s*, is Sansc. *snu*, fluere, stillare, (whence Germ. *schnee*, Eng. *snow*, &c.)

> *Germany.* ZNUUIA, 11th cent., now the SCHNEI.
> *Russia.* The ZNA or TZNA.

A derivative form is the Gael. and Ir. *snidh* or *snith*, to ooze through, distil, Obs. Gael. and Ir. *snuadh*, to flow, and *snuadh*, a river, whence I take the following. Förstemann refers to Old High German *snidan*, Modern German *schneiden*, to divide, in the sense of a boundary, which is a root suitable enough in itself, though I think it ought to yield the preference to the direct sense of water.

> *England.* The SNYTE. Leicestershire.
> *Germany.* SNEID(BACH), 8th cent., seems to be now called the Aue.

Germany. SMID(AHA), 9th cent., now the SCHMI-
DA, which joins the Danube. For
Snidaha ?

The form *snid* or *snith* introduces the form
nid or *nith*, and suggests the enquiry whether that may not also be a word signifying
water, Donaldson, (*Varronianus*), referring
to a word Nethuns, "found on a Tuscan
mirror over a figure manifestly intended for
Neptune," observes that "there can be little
doubt that *nethu* means water in the Tuscan
language." Assuming the correctness of the
premises, I think that this must be the case;
and that as the Naiades (water-nymphs),
contain the Greek *ναω*; as Nereus (a water-
god), contains the word *ner* before referred
to ; as Neptune contains the Greek *νίπτω*, in
each case involving the signification of water,
so Nethuns (= Neptunus) must contain a related word *neth* or *nethun* of the same meaning. Also that this word comes in its place
here, as a derivative of the root *nî*, and as a
corresponding form to the Celtic *snidh* or
snith.

There are, however, two other meanings
which might intermix in the following names;
the one is that suggested by Baxter, viz.,
Welsh *nyddu*, to turn or twist, in the sense
of tortuousness ; and the other is Old Norse
nidr, fremor, strepitus.

1. *England.* The NIDD. Yorkshire.
   Scotland. The NITH. Dumfriesshire.
   Wales. The NEATH. Glamorgan.
   France. The NIED. Joins the Sarre.
   Belgium. The NETHE. Joins the Ruppel.
   Germany. NIDA, 8th cent., now the NIDDA.
   The NETHE. Joins the Weser.
   Norway. The NIDA.
   Poland. The NIDDA.
   Greece. NEDA ant., now the Buzi in Elis.

2. *With the ending en.*
   Scotland. The NETHAN. Lesmahago.

3. *With the ending rn (see note p. 34.)*
   Germany. NITORNE, 9th cent., now the NIDDER.

There can hardly be a doubt that the
words *sar, sor, sur,* so widely spread in the
names of rivers, are to be traced to the Sansc.
sar, sri, to move, to go, *sru,* to flow, whence
saras, water, *sarit, sróta,* river. The Permic
and two kindred dialects of the Finnic class

have the simple form *sor* or *sur*, a river, and the Gaelic and Irish have the derived form *sruth*, to flow, *sroth*, *sruth*, river. In the names Sorg, Sark, Sarco, I rather take the guttural to have accrued.

1. *England.*   The SOAR.   Leicester.
The SARK, forms the boundary between England and Scotland.

France.   The SERRE.   Joins the Oise.

Germany.   SARAVUS ant., now the SAAR.
SORAHA, 8th cent., a small stream seemingly now unnamed.
SURA, 7th cent.   The SURE and the SUR.
The SORG.   Prussia.

Switzerland. The SARE and the SUR.

Norway.   The SURA.

Russia.   The SURA.   Joins the Volga.
The SVIR, falls into Lake Ladoga.

Lombardy.   The SERIO.   Joins the Adda.
The SERCHIO or SARCO.

Portugal.   The SORA.   Joins the Tagus.

Asia.   SERUS ant., now the Meinam.

Asia Minor. SARUS ant., now the Sihon.

India.   SARAYU* ant., now the Sardju.

---

* "One of the sacred rivers of India, a river mentioned in the Veda, and famous in the epic poems as the river of Ayodhyâ, one of the earliest capitals of India, the modern Oude."—*Max Müller, Science of Language.*

Armenia.	ARIUS* ant., now the Heri Rud.

2. *With the ending en.*

France.	The SERAN. Joins the Rhone.
	The SERAIN. Joins the Yonne.
Germany.	SORNA, 8th cent. The ZORN.
Switzerland.	The SUREN. Cant. Aargau.
Naples.	SARNUS ant. The SARNO.
Persia.	SARNIUS ant., now the Atrek.

The form *saras*, water, seems to be found in the following two names.

1. *With the ending en.*

France.	The SARSONNE. Dep. Corrèze.

2. *Compounded with wati = Goth, wato, water.*

India.	The SARASWATI, which still retains its ancient name.

And the Sansc. *sarit*, Gael. and Ir. *sroth*, *sruth*, a river, seem to be found in the following.

Ireland.	The SWORDS river near Dublin.
France.	The SARTHE. Joins the Mayenne.
Galicia.	The SERED. Joins the Dniester.
Moldavia.	The SERETH. Ant. Ararus.
Russia.	The SARAT(OVKA).† Gov. Saratov.

---

* I place this here on the authority of Max Müller, who, pointing out that the initial *h* in Persian corresponds with a Sanscrit *s*, thinks that the river Sarayu may have given the name to the river Arius or Heri, and to the country of Herat.

† This name seems formed at thrice—first Sarit—then ov, (perhaps *av* river)—lastly, the Slavish affix *ka*.

It would seem that the foregoing forms *sri, sru, srot,* sometimes take a phonetic *t,* and become *stri, stru, strot.* Thus one Celtic dialect, the Armorican, changes *sur* into *ster,* and another, the Cornish, changes *sruth* into *struth*—both words signifying a river. But indeed the natural tendency towards it is too obvious to require much comment. Hence we may take the names Stry and Streu. But is the form Stur from this source also ? Förstemann finds an etymon in Old High German *stur,* Old Norse *stôr,* great. This may obtain in the case of some of the rivers of Scandinavia, but is hardly suited for those of England and Italy, none of which are large. The root, moreover, seems too widely spread, if, as I suspect, it is this which forms the ending of many ancient names, as the Cayster, the Cestrus, the Alster, Elster, Ister, Danastris, &c. The Armorican *ster,* a river, seems to be the word most nearly concerned.

H

1. *The form stry, stru, stur.*

England. STURIUS (Ptolemy). The STOUR. There are six rivers of this name.

Germany. STROWA, 8th cent. The STREU.

Holstein. STURIA, 10th cent. The STÖR.

Italy. STURA, two rivers. STORAS (Strabo), now the ASTURA.

Aust.-Poland. The STRY. Joins the Dniester. The STYR. Joins the Pripet.

2. *The form struth.*

England. The STROUD. Gloucester. The STORT. Essex.

Germany. The UNSTRUT Förstemann places here, as far as the ending *strut* is concerned.

From the Sanscrit root *su*, liquere, come Sansc. *sava*, water, Old High German *sou*, Lat. *succus*, moisture, Gael. *sûgh*, a wave, &c.; (on the apparent resemblance between Sansc. *sava*, water, and Goth. *saivs*, sea, Diefenbach observes, we must not build). Hence I take to be the following; but a word very liable to intermix is Gael. *sogh*, tranquil; and where the character of stillness is very marked, I have taken them under that head.

1. *England.*   The Sow.   Warwickshire.
   Ireland.   The Suck.   Joins the Shannon.
   France.   The Save.   Joins the Garonne.
   Belgium.   Sabis, 1st cent. B.C., now the Sambre.
   Germany.   Savus ant.   The Save or Sau.
                The Söve.   Joins the Elbe.
   Russia.   The Seva.
   Italy.   The Savio.   Pont. States.
                The Sieve.   Joins the Arno.
2.   *With the ending en.*
   Italy.   The Savena or Saona.   Piedmont.
   Armenia.   The Sevan.   Lake.
3.   *With the ending er.*
   Ireland.   Severus ant.   The Suire.
   Germany.   Sevira, 9th cent.   The Zeyer.
   France.   The Sevre.   Two rivers.
   Spain.   Sucro ant.   The Xucar.
   Portugal. The Sabor.
4.   *With the ending rn (see note p. 34).*
   England.   Sabrina ant.   The Severn.
   France.   The Sevron.   Dep. Saône-et-Loire.
   Russ. Pol. The Savran(ka).   Gov. Podolia.
5.   *With the ending es.*
   Lombardy. The Savezo near Milano.

In the Sanscrit *mih*, to flow, to pour, Old Norse *miga*, scaturire, Anglo-Saxon *migan, mihan*, to water, Sansc. *maighas*, rain, Old Norse *migandi*, a torrent—("unde," says Haldorsen, "nomina propria multorum tor-

rentium"), Obs. Gael. and Ir. *machd*, a wave,
I find the root of the following. Most of the
names are no doubt from the Celtic, though
the traces of the root are more faint in that
tongue than in the Teutonic. This I take
to be the word, which in the forms *ma*, and
man or *men*, forms the ending of several
river-names.

1. *Scotland.*	The MAY.	Perthshire.
Ireland.	The MAIG and the MOY.	
Wales.	The MAY and the MAW.	
France.	The MAY.	
Siberia.	The MAIA.	Joins the Aldon.
India.	The MHYE.	Bombay.
2.	*With the ending en.*	
England.	The MAWN.	Notts.
	The MEON.	Hants. (Meôn eâ, *Cod. Dip.*)
Ireland.	The MAIN and the MOYNE.	
France.	The MAINE.	Two rivers.
Belgium.	The MEHAIGNE.	Joins the Scheldt.
Germany.	MOENUS ant.	The MAIN.
Sardinia.	The MAINA.	Joins the Po.
Siberia.	The MAIN.	Joins the Anadyr.
India.	The MEGNA.	Prov. Bengal.
	The MAHANUDDY—here ?	
3.	*With the ending er.*	
Italy.	The MAGRA.	Falls into the Gulf or Genoa.

4.   *With the ending el.*

England.   The MEAL. Shropshire.

Denmark.   The MIELE. Falls into the German Ocean.

5.   *With the ending st.**

Asia Minor. The MACESTUS. Joins the Rhyndacus.

From the root *mî*, to flow, come also Sansc. *mîras,* Lat. *mare,* Goth. *marei,* Ang.-Sax. *mêr,* Germ. *meer,* Welsh *mar, mor,* Gael. and Ir. *muir,* Slav. *morie,* &c., sea or lake. I should be more inclined however to derive most of the following from the cognate Sansc. *mærj,* to wash, to water, Lat. *mergo,* &c. Also, the Celtic *murg,* in the more definite sense of a morass, may come in for some of the forms.

1. *France.*   The MORGE. Dep. Isère.

Germany.   MARUS (Tacitus).   The MARCH, Slav. MOR(AVA).

               MUORA, 8th cent. The MUHR.

               MURRA, 10th cent. The MURR.

Belgium.   MURGA, 7th cent. The MURG.

               The MARK. Joins the Scheldt.

Switzerland. The MURG. Cant. Thurgau.

Sardinia.   The MORA. Div. Novara.

---

* See note p. 29.

Servia.	MARGUS ant. The MORAVA.
Italy.	The MARECCHIA. Pont. States— here ?
India.	The MERGUI—here ?
2.	*With the ending en.*
Ireland.	The MOURNE. Ulster.
Germany.	MARNE, 11th cent., now the MARE. MERINA, 11th cent. The MÖRN.
3.	*With the ending es.*
England.	The MERSEY. Lancashire.
Germany.	MUORIZA, 10th cent. The MURZ.
Dacia.	MARISUS ant. The MAROSCH.
Phrygia.	MARSYAS ant.

Another form of Sansc. *marj*, to wet, to wash, is *masj*, whence I take the following.

Ireland.	MASK, a lake in Connaught.
Russia.	The MOSK(VA), by Moscow, to which it gives the name.

From the Sanscrit *vag* or *vah*, to move, comes *vahas*, course, flux, current, cognate with which are Goth. *wegs*, Germ. *woge*, Eng. *wave*, &c.   An allied Celtic word is found as the ending of many British river-names, as the Conway, the Medway, the Muthvey, the Elwy, &c.   Hence I take to be the following, in the sense of water or river.

1. *England.*   The WEY.   Dorset.

                The WEY.   Surrey.

    Hungary.   The WAAG.   Joins the Danube.

    Russia.   The VAGA.   Joins the Dwina.

                The VAGAI and the VAKH in Siberia.

    India.   The VAYAH.   Madras.

2.   *With the ending en.*

    England.   The WAVENEY.   Norf. and Suffolk.

3.   *With the ending er.*

    England.   The WAVER.   Cumberland.

4.   *With the ending el.*

    Netherlands. VAHALIS, 1st cent. B.C. The WAAL.

5.   *With the ending es = Sansc. vahas ?*

    France.   VOGESUS ant.   The VOSGES.

An allied form to the above is found in Sansc. *vi, vic,* to move, Lat. *via,* &c., and to which I put the following.

1. *England.*   The WYE.   Monmouthshire.

    Scotland.   The WICK.   Caithness.

    France.   The VIE.   Two rivers.

    Russia.   The VIG.   Forms lake VIGO.

2.   *With the ending en.*

    France.   VIGENNA ant.   The VIENNE.

    Germany.   The WIEN, which gives the name to Vienna, (Germ. Wien).

3.   *With the ending er.*

    Switzerland. The WIGGER.   Cant. Lucerne.

    France.   The VEGRE.   Dep. Sarthe.

                The VIAUR—probably here.

  Poland. The WEGIER(KA).
  India.   The VEGIAUR, Madras—here ?

Formed on the root *vi*, to move, is probably also the Sansc. *vip* or *vaip*, to move, to agitate, Latin *vibrare*, perhaps *vivere*, Old Norse *vippa*, *vipra*, gyrare, Eng. *viper*, &c. I cannot trace in the following the sense of rapidity, which we might suspect from the root. Nor yet with sufficient distinctness the sense of tortuousness, so strongly brought out in some of its derivatives.

 1.   *With the ending er.*
  England. The WEAVER. Cheshire.
      The VEVER. Devonshire.
  Germany. WIPPERA, 10th cent. The WIPPER
      (two rivers), and the WUPPER.
 2.   *With the ending es.*
  India. VIPASA, the Sanscrit name of the
      Beas.
 Switzerland. VIBSICUS ant. (properly Vibissus ?)
      The VEVEYSE by Vevay.

From the root *vip*, to move, taking the prefix *s*, is formed *swip*, which I have dealt with in the next chapter.

In the Sansc. *par*, to move, we find the root of Gael. *beathra* (pronounced *bcara*),

Old Celt. *ber*, water, Pers. *baran*, rain, &c., to which I place the following.

1. *England.*   The BERE. Dorset.
   Ireland.   BARGUS (Ptolemy). The BARROW.
   France.   The BAR. Dep. Ardennes.
      The BERRE. Dep. Aude.
   Germany.   The BAHR, the BEHR, the BEHRE, the PAAR.
2.   *With the ending en.*
   Bohemia.   The 'BERAUN near Prague.
   India.   The BEHRUN.
   Russia.   The PERNAU. Gulf of Riga.

From the Sansc. *plu*, to flow, Lat. *pluo* and *fluo*, come Sansc. *plavas*, flux, Lat. *pluvia* and *flluvius*, Gr. πλυνω, lavo, Ang.-Sax. *flôwe*, *flum*, Lat. *flumen*, river, &c. Hence we get the following.

1. *Germany.*   The PLAU, river and lake.* Mecklenburg-Schwerin.
   Holland.   FLEVO, 1st cent. The Zuiderzee, the outlet of which, between Vlieland and Schelling, is still called VLIE.
   Aust. Italy.   PLAVIS ant. The PIAVE, falls into the Adriatic.
2.   *With the ending en.*
   France.   The PLAINE. Joins the Meurthe.

---

* In the more special sense of lake, which, it will be observed, is frequent in this group, is the Suio-Lapp. *pluewe.*

I

Germany.	The PLONE. Joins the Haff.
	The PLAN-SEE, a lake in the Tyrol.
Holstein.	PLOEN. A lake.
Poland.	The PLONNA. Prov. Plock.

From the above root come also the following, which compare with Sansc. *plavas,* Mid. High Germ. *vlieze,* Mod. Germ. *fliess,* Old Fries. *flêt,* Old Norse *fliot,* stream. And I think that some at least of this group are German.

1.	*England.*	The FLEET. Joins the Trent.
		The FLEET, now called the Fleetditch in London.
	Scotland.	The FLEET. Kirkcudbright
	Germany.	BLEISA, 10th cent. The PLEISSE.
	Holland.	FLIETA, 9th cent. The VLIET.
	Russia.	The PLIUSA. Gulf of Finland.
2.		*With the ending en.*
	Germany.	FLIEDINA, 8th cent. The FLIEDEN.
		The FLIETN(ITZ). Pruss. Pom.
3.		*With the ending st.*
	Holland.	The VLIEST.
	Greece.	PLEISTUS ant., near Delphi.

There are two more forms from the same root, the former of which we may refer to the Irish and Gael. *fluisg,* a flushing or flowing. The latter shows a form nearest to the

Ang.-Sax. and Old High Germ. *flum*, Lat. *flumen*, though I think that the names must be rather Celtic.

1. *Ireland.* The FLISK. Falls into the Lake of Killarney.
   Germany. The PLEISKE. Joins the Oder.
2. *England.* The PLYM, by Plymouth.
   Scotland. The PALME, by Palmton.
   Siberia. The PELYM. Gov. Tobolsk.

From the Sansc. *gam*, to go, is derived, according to Bopp and Monier Williams, the name of the Ganges, in Sanscrit Gangâ. The word is in fact the same as the Scotch "gang," which seems to be derived more immediately from the Old Norse *ganga*. In the sense of "that which goes," the Hindostanee has formed *gung*, a river, found in the names of the Ramgunga, the Kishengunga, the Chittagong, and other rivers of India. The same ending is found by Förstemann in the old names of one or two German rivers, as the Leo near Salzburg, which in the 10th cent. was called the LIUGANGA. Another name for the Ganges is the Pada, for which

Hindoo ingenuity has sought an origin in the myth of its rising from the foot of Vishnoo. But as *pad* and *gam* in Sanscrit have both the same meaning, viz., to go, I am inclined to suggest that the two names Ganga and Pada may simply be synonymes of each other.

1. *India.* The GANGES. Sanscrit GANGA.
   The GINGY. Pondicherry.
   Russia. The KHANK(OVA). Joins the Don.
2.   *With the ending et.*
   Greece. GANGITUS ant., in Macedonia.

The Sansc. verb *gam*, to go, along with its allied forms, is formed on a simpler verb *gâ*, of the same meaning. To this I put the following.

1. *Holland.* The GOUW. Joins the Yssel.
   Persia. CHOES or CHO(ASPES)* ant.
2.   *With the ending en.*
   Germany. GEWIN(AHA), 9th cent., now the JAHN(BACH).
3.   *Compounded with ster, river.*
   Asia Minor. The CAYSTER and CESTRUS—here ?

---

* The word *asp* comes before us in some other river-names, but respecting its etymology I am quite in the dark. From the way in which it occurs in the above, in the Zari(aspis), and in the Hyd(aspes), it seems rather likely to have the meaning of water or river.

The Sansc. *ikh*, to move, must, I think, contain the root of the following, though I find no derivatives in any sense nearer to that of water or river.

1. *Russia.* The IK. Two rivers.
2. *With the ending en.*
   England. ICENA (*Cod. Dip*). The ITCHEN.
   France. ICAUNA ant. The IONNE.
3. *With the ending el.*
   Moravia. The IGLA or IGL(AWA).
   France. The ECOLLE. Dep. Seine-et-Oise.

From the Sansc. *dravas*, flowing, are derived, according to Bopp, the Drave and the Trave. The root-verb is, I presume, *drâ*, to move. Hence I have suggested, p. 37, may be the Welsh *dwr*, water.

1. *Scotland.* The TARF, several small rivers—here?
   Germany. DRAVUS, 1st cent. The DRAVE, Germ. DRAU.
   Italy. The TREBBIA. Joins the Po.
2. *With the ending en.*
   Germany. TRAVENA, 10th cent., now the TRAVE.
   TREWINA, 9th cent. The DRAN.
   DRONA, 9th cent. The DRONE.
   TRUNA, 7th cent. The TRAUN.
   France. The DRONNE. Joins the Isle.

In the Sansc. *dram*, to move, to run, Gr. δρέμω, whence *dromedary*, &c., is to be found the root of the following. But *dram*, as I take it, is an interchanged form with the preceding *drav*, as *amon* = *avon*, &c., *ante*.

1.	*Scotland.*	The TROME and the TRUIM. Inverness.
	France.	The DROME and the DARME.
	Belgium.	The DURME.
	Germany.	The DARM, by Darmstadt.
2.		*With the ending en.*
	Norway.	The DRAMMEN. Christiania Fjord.

Another word of the same meaning as the last, and perhaps allied in its root, is Sansc. *trag*, to run, Gr. τρέχω, Goth. *thragjan*. It will be observed that the above Greek verb mixes up in its tenses with the obsolete verb δρέμω of the preceding group. In all these words signifying to run there may be something of rapidity, though I am not able to remove them out of this category.

1.	*France.*	The DRAC. Joins the Isère.
	Prussia.	The DRAGE.
	Greece.	TRAGUS ant.
	Italy.	The TREJA. Joins the Tiber.

2.   *With the ending en.*
 Sicily. The TRACHINO. Joins the Simeto.

The Sansc. *il*, to move, Gr. ἔίλω, Old High Germ. *ilen*, Swed. *ila*, Mod. Germ. *eilen*, to hasten, Fr. *aller*, &c., is a very widely spread root in river-names.

1. *England.* The ILE. Somerset.
     The ALLOW. Northumberland.
 France. The ILL, the ILLE, and the ELLÉ.
 Germany. ILLA, 9th cent. The ILL.
     IL(AHA), 11th cent. The IL(ACH).
     The ALLE. Prussia.
 Italy.   ALLIA ant., near Rome.
2.   *With the ending en.*
 England. ALAUNUS (Ptolemy). Perhaps the
     Axe.
     The ALNE, two rivers.
     The ELLEN. Cumberland.
 Scotland. The ALLAN, two rivers.
 Ireland. The ILEN. Cork.
 France. The AULNE. Dep. Finistère.
3.   *With the ending er.*
 Germany. ALARA, 8th cent. The ALLER.
     ILARA, 10th cent. The ILLER.
 Piedmont. The ELLERO.

From the above root *al* or *il*, to move, to go, I take to be the Gael. *ald* or *alt*, a stream, (an older form of which, according to Arm-

strong, is *aled*) ; and the Old Norse *allda*,
Finnish *aalto*, a wave, billow. As an ending
this word is found in the NAGOLD of Germany (ant. NAGALTA), and in the HERAULT
of France, Dep. Herault. Förstemann makes
the former word *nagalt*, and remarks on it
as "unexplained." It seems to me to be a
compound word, of which the former part is
probably to be found in the root *nig* or *nî*,
p. 47.

1. *England.*	The ALDE.	Suffolk.
	The ALT.	Lancashire.
France.	OLTIS ant., now the Lot.	
Germany.	The ELD.	Mecklenburg-Schwerin.
Spain.	The ELDA.	
Russia.	The ALTA.	Gov. Poltova.
2.	*With the ending en.*	
Germany.	ALDENA, 11th cent., now the Olle.	
Norway.	The ALTEN.	
Siberia.	The ALDAN.	Joins the Lena.

Also from the root *al* or *il*, to move, I take
to be the Old Norse *elfa*, Dan. *elv*, Swed. *elf*,
a river. The river Ἄλπις mentioned in Herodotus is supposed by Mannert to be the Inn
by Innsbrück. I think the able Editor of

Smith's Ancient Geography has scarcely sufficient ground for his supposition that Herodotus, in quoting the Alpis and Carpis as
rivers, confounded them with the names of
mountains. The former, it will be seen, is
an appellative for a river ; the latter is found
in the name Carpino, of an affluent of the
Tiber, and might be from the Celt. *garbh*,
violent ; a High Germ. element, for instance,
would make *garbh* into *carp*. But indeed
the form *carp* is that which comes nearest
to the original root, if I am correct in supposing it to be the the Sansc. *karp*, Lat.
carpo, in the sense of violent action. In the
following list I should be inclined to take
the names Alapa, Elaver, and Ilavla, as nearest to the original form.

1. *Germany.* ALBIS, 1st cent. The ELBE. Also
the ALB in Baden, and the ALF in
Pomerania.
ALPIS (Herodotus), perhaps the Inn.
ALAPA, 8th cent., now the Wölpe.
The AUPE. Joins the Elbe.

France. ALBA ant., now the AUBE.

K

	The AUVE. Dep. Marne.
	The HELPE. ¡ Joins the Sambre.
Greece.	ALPHEUS ant., now the Rufio—here?
2.	*With the ending en.*
Scotland.	The ELVAN. Joins the Clyde.
Germany.	ALBANA, 8th cent., now the ALBEN.
Tuscany.	ALBINIA ant. The ALBEGNA.
3.	*With the ending er.*
France.	ELAVER ant., now the Allier.
4.	*With the ending el.*
Germany.	ALBLA, 11th cent., not identified.
Italy.	ALBULA, the ancient name of the Tiber.
Russia.	The ILAVLA. Joins the Don.

Förstemann seems to me to be right in his conjecture that the forms *alis, els, ils,* are also extensions of the root *al, el, il.* We see the same form in Gr. ἐλισσω, an extension of ἐιλω, and having just the same meaning of verso, volvo. Indeed I think that this word, which we find specially applied to rivers, is the one most concerned in the following names, two of which, it will be seen moreover, belong to Greece. Hence may perhaps be derived the name of the Elysii, (wanderers?) a German tribe mentioned in Tacitus.

And through them, of many names of men, as the Saxon Alusa and Elesa, down to our own family names Alice and Ellice.*

1. *France.* The ALISE.
 Germany. ELZA, 10th cent., now the ELZ.
     ILSA ant., now the ILSE.
     The ALASS. Falls into the Gulf of Riga.
 Greece. ILISSUS ant., still retains its name.
 Asia Minor. HALYS ant., now the Kizil-Irmak.

2.   *With the ending en*
 Germany. ELISON, 3rd cent., now the Lise.
 Belgium. ALISNA, 7th cent., not identified.
 Greece. ELLISON or HELISSON ant.

3.   *With the ending es.*
 Germany. ALZISSA, 9th cent., now the ALZ.
     ILZISA, 11th cent., now the ILZ.

The root *sal* Förstemann takes to be Celtic, and to mean salt water. No doubt saltness is a characteristic which would naturally give a name to a river. So it does in the case of the "Salt River" in the U.S., and of the Salza in the Salzkammergut. But I can

---

* Also ALLISON and ELLISON, which may be either patronymic forms in *son ;* or formed with the ending in *en*, like the above river-names. For the names of rivers, and the ancient names of men, in many points run parallel to each other.

hardly think that all the many rivers called the SAALE are salt, and I am inclined to go deeper for the meaning. The Sansc. has *sal*, to move, whence *salan*, water. The first meaning then seems to be water—applied to the sea as *the* water—and then to salt as derived from the sea. So that when the Gr. ἅλς, the Old Norse *salt*, and the Gael. *sal*, all mean both salt, and also the sea, the latter may be the original sense. From the above root, *sal*, to move, the Lat. forms both *salire* and *saltare*, as from the same root come *sal* and *salt*. I take the root *sal* then in river-names to mean, at least in some cases, water. In one or two instances the sense of saltness comes before us as a known quality, and in such case I have taken the names elsewhere. But failing the proper proof, which would be that of tasting, I must leave the others where they stand.

1. *Germany.*   SALA, 1st cent.   Five rivers called the SAALE.

SALIA, 8th cent.   The SEILLE.

France.   The SELLÉ.   Two rivers.

Russia.	The SAL. Joins the Don.	
Spain.	SALO ant., now the XALON.	

2.   *With the ending en=Sansc. salan, water ?*

Ireland.	The SLAAN and the SLANEY.
France.	The SELUNE. Dep. Manche.

It is possible that the root *als, ils,* found in the name of several rivers, as the ALZ, ELZ, ILSE, may be a transposition of the above, just as Gr. ἁλς = Lat. *sal.* But upon the whole I have thought another derivation better, and have included them in a preceding group.

From the Sansc. *var* or *vars*, to bedew, moisten, whence *var*, water, *varsas*, rain, Gr. ἑρση, dew, Gael. and Ir. *uaran*, fresh water, I get the following, dividing them into the two forms, *var* and *vars*.

The form var.

1. *England.*	The VER. Herts.	
France.	VIRIA ant. The VIRE.	
Germany.	The WERRE in Thuringia.	

2.   *With the ending en.*

Germany.	WARINNA, 8th cent. The WERN.	
	The WARN(AU). Mecklenburg-Schweren.	

Naples.   VARANO,* a lagoon on the Adriatic
                   shore.

                   The form vars.
1. *England.*   The WORSE.   Shropshire.
France.   The OURCE.   Joins the Seine.
Germany.   The WERS.   Joins the EMS.
Italy.   ARSIA ant.—here ?
                   VARESE.   Lake in Lombardy.
Persia.   AROSIS ant., now the Tab—here ?
Armenia.   †ARAXES ant., now the ARAS—here ?
2.   *With the ending en.*
Germany.   URSENA, 8th cent., now the OERTZE.
Asia Minor.   ORSINUS ant., now the Hagisik—
                   here ?

3.   *With the ending el.*
Germany.   URSELA, 8th cent.   The URSEL.
                   HÖRSEL.   Joins the Werre.

In the above Sansc. *var*, to moisten, to
water, is contained, as I take it, the root of
the Finnic *wirta*, a river, the only appellative I can find for the following.

1. *Germany.*   WERT(AHA), 10th cent., now the
                   WERT(ACH).

---

* Following strictly the above Celt. word *uaran*, this might be
"Fresh-water Bay."

† The Araxes of Herodotus, observes the Editor of Smith's Ancient
Geography, "cannot be identified with any single river : the name was
probably an appellative for a river, and was applied, like our Avon, to
several streams, which Herodotus supposed to be identical." Araxes I
take to be a Græcism, and the Mod. name Aras to show the proper form.

Poland.	The WARTA.	Joins the Oder.
Denmark.	The VARDE.	Prov. Jütland.
India.	The WURDAH.	Joins the Goda-avery.

2.   *With the ending en.*

France.	The VERDON.	Dep. Var.

3.   *With the ending er.*

Ireland.	The VARTREY.	Wicklow.
France.	The VARDRE.	
Europ. Turkey.	The VARDAR, ant. Axius.	

The following names have been generally supposed to be derived from Welsh *cledd* or *cleddeu,* sword, and to be applied metaphorically to a river. But I think it will be seen from the Sansc. *klid,* to water, whence *klaidan,* flux, Gr. κλύδων, fluctus, unda, Ang.-Sax. *glade,* a river, brook, that the meaning of water lies at the very bottom of the word. Perhaps, however, as the senses of a running stream and of a sharp point often run parallel to each other, there may be in this case a relationship between them.

1. *Scotland.*	The CLYDE.	(CLOTA, Ptolemy.)
Wales.	The CLOYD, the CLWYD, and the CLEDDEU.	

Ireland.	The GLYDE.
Greece.	CLADEUS ant.—here ?
Umbria.	CLIT(UMNUS)* ant.—here ?
2.	*With the ending en.*
Germany.	The KLODN(ITZ). Pruss. Silesia.
3.	*With the ending er.*
Greece.	The CLITORA in Arcadia, on which stood the ancient Clitorium.
Asia Min.	CLUDROS ant., in Caria.

There are two Sanscrit roots from which
the word *ag, ang, ing,* in river-names might
be deduced.   One is the verb *ag* or *aj,* to
move, whence *anjas,* movement, (or the verb
ac or *anc,* to traverse), and the other is the
verb *ag* or *ang,* to contract, whence Latin
anguis, snake, *anguilla,* eel, Eng. *angle,* &c.
The sense then might be either the ordinary
one of motion, the root-meaning of most
river names, or it might be the special sense
of tortuousness.   But as the only appellative
I can find is the word *anger,* a river, in the
Tcheremissian dialect of the Finnic (Bona-
parte polyglott), I think it safer to follow
the most common sense, though the other

---

* Containing the Latin *amnis,* river, or only a euphonic form of
Clitunnus ? See Garumna, p. 13.

may not improbably intermix. The derivation of Mone, from Welsh *eog*, salmon, I do
not think of.

1.		*With the ending en.*
	Germany.	ANKIN(AHA), 8th cent., now the ECKN(ACH).
	France.	The INGON. Dep. Somme.
2.		*With the ending er.*
	England.	The ANKER. Leicestershire.
	Germany.	ACKARA, 10th cent. The AGGER. AGARA, 8th cent. The EGER. The ANGERAP (*ap*, water), Prussia.
	Siberia.	The ANGERA.
	Italy.	ACARIS ant. The AGRI.
	Servia ?	ANGRUS (Herodotus).
	India.	The AGHOR—here ?
3.		*With the ending el.*
	Germany.	The ANGEL, three rivers (Baden, Westphalia, and Bohemia).
	Russia.	The INGUL. Joins the Bug.
4.		*With the ending st.*
	Germany.	AGASTA,* 8th cent., now the AISS.

From the Sansc. *pî*, to drink, also to give
to drink, to water, Gr. πιω, πινω, we may get
a form *pin* in river-names.

1. *Germany.* The PEEN in Prussia.

---

* I think that in this, as probably in some other cases, *st* is only a
phonetic form of *ss*, and that the Mod. name *Aiss* points truly to the
ancient form as *Agass*, see note, p. 29.

Holstein.	The PINAU.	Joins the Elbe.
Hungary.	The PINA.	Joins the Pripet.
	The PINKA—here ?*	
Russia.	The PIANA.	Joins the Volga.
	The PINE(GA).	Joins the Dwina.
India.	The BINOA.	Joins the Beas.
Greece.	PENEUS, ant.	Two rivers—here ?
2.	*With the ending en.*	
Siberia.	The PENJINA.	
3.	*With the ending er.*	
India.	The PENNAR.	Madras.
4.	*With the ending es.*	
Russia.	The PENZA.	Joins the Sura.

From the above Sansc. *pi* we may also derive the form *pid*. The only appellative I find, (if it can be called one), is the Ang.-Sax. *pidele*, a thin stream, given by Kemble in the glossary to the *Cod. Dip.;* and hence the name PIDDLE, of several small streams. The only name I find in the simple form, and that uncertain, is the PINDUS of Greece. Then there is a form *peder*, which seems to be from a definite word, and not from the simple suffix *er*.

---

* I should without hesitation have taken the PINKA, as well as the Russian PINEGA, to be from this root, with the Slavonic affix *ga* or *ka*. But the English river PENK in Staffordshire introduces an element of doubt. It may, however, also be from this root, with the ending *ick* common in the rivers of Scotland. See p. 25.

1. *England.*   The PEDDER.   Somerset.
    Greece.   PYDARAS ant. Thrace.
    India.   The PINDAR—here ?

2.   *With the ending en.*
   Scotland.   The PITREN(ICK), a small stream in Lanarkshire.

3.   *With the ending el.*
   England.   The PETTERIL in Cumberland.

4.   *With the ending et.*
   England.   PÊDREDE (*Cod. Dip.*) Now the PARRET.

Also from the Sansc. root *pi*, to drink, to water, we get the form *bib* or *pip*, as found in Lat. *bibo*, and in Sansc. *pipâsas*, toper. Here also in the simple form I only find one name—the BEUVE in France, Dep. Gironde. In the form *biber* there are many names, particularly in Germany. Graff (*Sprachschatz*), seems to refer the word to *biber*, beaver, but Förstemann, with more reason, as I think, suggests a lost word for water or river.

1. *England.*   The PEVER.   Cheshire.
   Scotland.   The PEFFER.   Ross-shire.
   France.   The BIÈVRE.   Joins the Seine.

> *Germany.* BIBER(AHA), 7th cent. The BEVER,
> the BIBRA, the PEBB(ACH), and the
> BIBER(BACH).

2.   *With the ending en.*
> *Germany.* BIVERAN, 8th cent., now the BEVER.
> *France.* The BEUVRON. Dep. Nièvre.

Perhaps also from the root *pi* we may
derive the Ir. *buinn*, river, *bual, biol,* water.
From the former Mr. Charnock derives the
name of the Boyne, a derivation which I
think suitable, even if we take the ancient
form Buvinda, *(Zeuss, Gramm. Celt.,)* which
might be more properly Buvinna, as Gironde
for Garonne in France. For the Bunaha in
Germany, the Old Norse *buna*, scaturire,
might also be suggested.

> *Ireland.* The BOYNE.
> *Germany.* BUN(AHA), 9th cent., now the BAUN-
> (ACH).

From the Ir. *biol, buol,* I derive the following, keeping out the rivers of the Slavonic
districts, which may be referred to the Slav.
biala, white.

> 1. *England.* The BEELA. Westmoreland.

Ireland.	The BOYLE, of which, according to O'Brien, the Irish form is BUIL.	
France.	The BOL(BEC). Dep. Seine-Inf.	
Germany.	BOLL(AHA) ant. Not identified.	
Asia Minor.	BILLÆUS ant., now the Filyas.	

2.		*With the ending er.*
	Germany.	The BUHLER. Wurtemburg.
	Russia.	The BULLER.

3.		*With the ending et.*
	Germany.	The BULLOT. Baden.
	Russia.	The POLOTA. Joins the Dwina.

A very obscure root in river-names is *gog* or *cock*. The only appellatives I find are in the Celtic, viz., Gael. *caochan*, a small stream, Arm. *goagen*, wave; unless we think also of the word *jokk*, *joggi*, which in the Finnic dialects signifies a river; and in that case the most probable root would be the Sansc. *yug*, to gush forth. To the river Coquet, in Northumberland, something of a sacred character seems to have been ascribed; an altar having been discovered bearing the inscription "Deo Cocidi," and supposed to have been dedicated to the genius of that river. Again, we are reminded of the Cocytus in

Greece, a tributary of the river Acheron, invested with so many mysterious terrors as supposed to be under the dominion of the King of Hades. Possibly, however, it might only be the similarity, or identity, of the names which transferred to the one something of the superstitious reverence paid to the other. At all events, I can find nothing in the etymology to bear out such a meaning.

1. *England.*		Cocbrôc (*Cod. Dip.*) This would seem to have probably been a small stream called Cock, to which, as in many other cases, the Saxons added the word brook.
2. *Germany.*		Cochin(aha), 8th cent., now the Kocher.*
3.		*With the ending er.*
	England.	The Cocker. Cumberland. The Coker. Lancashire.
	India.	The Kohary—here?
4.		*With the ending el.*
	Transylvania.	The Kokel, two rivers.

---

* This river seems also to have been called anciently Chockara.

England.	COCKLEY-BECK.*	Cumberland.
Germany.	CHUCHILIBACH, now Kuchel-bach.

5.   *With the ending et.*

England.	The COQUET.	Northumberland.
Greece.	COCYTUS ant., now the Vuvo.

6.   *In a compound form.*

England.   The CUCKMARE, Sussex, with the word mar, p. 61.

From the Sansc. *mïd*, to soften, to melt, (perhaps formed on the root *mi,* p. 59), come Sansc. *miditas*,. fluid, Lat. *madidus*,. wet. Herein seems a sufficient root for river-names, but there is another which is apt to intermix, Sansc. *math*, to move, whence, I take it, and not from the former, is Old Norse *môda*, a river. I separate a form *med* or *mid*, in which the sense of *medius*,. and also that of *mitis*, is in some cases clearly brought out ; and another, *muth* or *muot*, which, though from the same root, as I take it, as *môda*, a river, (*math*, to move), has more evidently the sense of speed.

---

* Here also, as in the case of the German Chuchilibach, and the Cocbrôc before noted, the ending beck (=brook), seems to have been added to the original name. Chuchilibach appears as the name of a place, but I apprehend that the word implies a stream of the same name.

1. *Germany.*   MOTA, 8th cent., now the MEDE or MEHE.

2.   *With the ending er.*
England.   The MADDER.   Wiltshire.
Germany.   MATRA, 8th cent., now the MODER.
Italy.   METAURUS ant., the METAURO—here?

3.   *With the ending ern.*
France.   MATRŎNA* ant., now the Marne.
Italy.   MATRINUS ant. in Picenum.

4.   *With the ending el.*
Germany.   The MADEL.

The only appellative for a river which I find derived from its sound is the Sanscrit *nadi,* Hind. *nuddy,* from *nad,* sonare. Whether the following names should come in here may be uncertain; I can find no links between them and the Sanscrit; perhaps the root *nid,* p. 54, may be suitable.

1. *France.*   NODA ant., now the Noain.

2.   *With the ending er.*
England.   The NODDER.   (Noddre, *Cod. Dip.*)
Hungary.   The NEUTRA.   Joins the Danube.

3.   *With the ending es.*
Venetia.   NATISO ant., now the NATISONE.

---

* I think that these quantities, so far as they are derived from the Latin poets, should be accepted with some reserve. Unless more self-denying than most of their craft, I fear that they would hardly let a Gallic river stand in the way of a lively dactyl.

The only words I can find at all bearing upon the following river-names are the Serv. *jezor*, Bohem. and Illyr. *jezero*, lake, wherein may probably lie a word *jez*, signifying water. But respecting its etymology I am entirely in the dark.

1. *Germany.*   Jaz(AHA), 8th cent., now the Joss.
                Jez(AWA), 11th cent., a brook near Lobenstein.
                The Jetza.   Joins the Elbe.
                The Jess(AVA).   Joins the Danube.

2.   *With the ending er.*
   Russia.   The Jisdra.   Joins the Oka.

3.   *Compounded with main, river.*
   Russia.   The Jesmen.   Gov. Tchnerigov.

Another word, of which the belongings are not clearly to be traced, is the Armorican *houl, houlen,* unda, to which we may put the following.

1. *England.*   The Hull.   Joins the Humber.
   Finland.   The Ullea.   Gulf of Bóthnia.
   Spain.   The Ulla in Galicia.

2.   *Compounded with ster, river.*
   Germany.   Ulstra, 9th cent., now the Ulster.

In the Irish and Obs. Gael. *dothar*, water, Welsh *diod*, drink, *diota*, to tipple—with

M

which we may perhaps also connect the Lapp. *dadno,* river, Albanian δέτ, sea, and Rhæt. *dutg,* torrent, we may find the root of the following.

1. *Germany.* The DUYTE. Joins the Hase.
     The DUDE, a small stream in Prussia.

2.   *With the ending en.*
 England. The DUDDON. Lake district.

3.   *With the ending er.*
 Ireland. The DODDER.

4.   *Compounded with mal.* *
 Germany. DUTHMALA, 8th cent., now the DOMMEL.

From the Welsh *wyl,* Ang.-Sax. *wyllan,* Eng. *well,* to flow or gush, (Sansc. *vail,* to move ?), we got the following.

1. *England.* The WILLY. Wiltshire.
 Denmark. The VEILE, in Jutland.
 Norway. The VILLA.
 Russia. The VEL. Joins the Vaga.
     The VILIA. Joins the Niemen.
     The VILIU, (Siberia). Joins the Lena.

2.   *With the ending en.*
 England. The WELLAND, (properly Wellan ?)
 Russia. The VILNA. Gov. Minsk.

---

* I do not know any other instance of this ending in river names, but I take it to be, like *man* or *main,* an extension of *may,* and to signify water or river.

Italy.	The VELINO.	Joins the Nera.
3.	*With the ending er.*	
India.	The VELLAUR, Madras—here ?	
4.	*With the ending s.*	
Germany.	The VILS, two rivers in Bavaria.	
	The WELSE.	Joins the Oder.
Spain.	The VELEZ.	Prov. Malaga.

A word which appears to have the meaning of water or river, but respecting the etymology of which I am quite ignorant, is *asop* or *asp*. That it has the above meaning I infer only from finding it as the second part of the word in the ancient river-names Cho-(aspes), Hyd(aspes), and Zari(aspis). In an independent form it occurs in the following. Lhuyd, (in the appendix to Baxter's glossary), referring to Hespin as the name of sundry small streams in Wales, derives it from *hespin*, a sheep that yields no milk, because these streams are almost dry in summer. This derivation is unquestionably false so far as this, that the two words are merely derived from the same origin, viz., Welsh *hesp* or *hysp*, dry, barren. But whether this word has anything to do with the following names is doubtful ; it seems at any rate unsuitable for the large rivers, such as the Hydaspes, (the Jhylum of the Punjaub). From the derivation of Mone, who finds in Isper, as in Wipper, p. 64, a word *per*, mountain, I entirely dissent.

1.	*France.*	The ASPE. Basses—Pyrenees.
	Germany.	HESAPA ant., now the HESPER.
	Greece.	ASOPUS ant. Two rivers.
2.		*With the ending er.*
	Germany.	ISPERA, 10th cent. The ISPER.

CHAPTER V.

THAT WHICH RUNS RAPIDLY, FLOWS GENTLY, OR SPREADS WIDELY.

In the preceding chapter I have included the words from which I have not been able to extract any other sense than that of water. As I have before mentioned, it is probable that in some instances there may be fine shades of difference which would remove them out of that category, but whenever I have thought to have got upon the trace of another meaning, something has in each case turned up to disappoint the conditions.

In the present chapter, which comprehends the words which describe a river as that which runs rapidly, that which flows gently, that which spreads widely, there may still in some cases be something of an appellative sense, because there may be a general word to denote a rapid, a smooth, or a spreading stream.

Among the rivers noted for their rapidity is the Rhone. This is the characteristic remarked by all the Latin poets—

> Testis Arar, Rhodanusque celer, magnusque Garumna.
> *Tibullus.*

> Qua Rhodanus raptim velocibus undis
> In mare fert Ararim.
> *Silv. Ital.*

> Præcipitis Rhodani sic intercisa fluentis.
> *Ausonius.*

I think that Donaldson and Mone are unquestionably wrong in making the name of this river Rho-dan-us, from a word *dan*, water. Still more unreasonable is a derivation in the *Cod. Vind.*, from *roth*, violent, and *dan*, Celt. and Hebr. a judge! On this Zeuss (*Gramm. Celt.*) remarks—"The syllable *an* of the word Rhodanus is without doubt only derivative, and we have nothing here to do with a judge ; nevertheless the meaning violent (currens, rapidus,) is not to be impugned." The word in question seems to be found in Welsh *rhedu*, to run, to race, Gael. *roth*, a wheel, &c. But there is a word of opposite meaning, Gael. *reidh*, smooth, which is liable to intermix. Also the Germ. *roth*, red, may come in, though I do not think that Förstemann has reason in placing all the German rivers to it.

<table>
<tr><td>1. England.</td><td>The ROTHA. Lake district.</td></tr>
<tr><td>Germany.</td><td>ROT(AHA), 8th cent. The ROTH, two rivers, the ROTT, three rivers, the ROD(AU), the ROD(ACH), and the ROTT(ACH), all seem to have had the same ancient name.</td></tr>
<tr><td></td><td>RAD(AHA) ant., now the ROD(ACH).</td></tr>
<tr><td>Holland.</td><td>The ROTTE, by Rotterdam.</td></tr>
<tr><td>Asia Min.</td><td>RHODIUS ant.* Mysia.</td></tr>
<tr><td>2.</td><td>With the ending en.</td></tr>
<tr><td>England.</td><td>The RODDEN. Shropshire.</td></tr>
<tr><td>France.</td><td>RHODĂNUS ant., now the RHONE.</td></tr>
<tr><td>Germany.</td><td>The ROTHAINE near Strassburg, seems to have been formerly ROT(AHA).</td></tr>
<tr><td>3.</td><td>With the ending ent.†</td></tr>
<tr><td>Germany.</td><td>RADANTIA, 8th cent., now the RED-NITZ.</td></tr>
</table>

---

* This, one of the Homeric rivers, was not identified in the time of Pliny.

† Perhaps formed from *et* by a phonetic *n*. So the Eamont in Cumberland seems to have been called in the time of Leland the Eamot.

4.   *With the ending er.*
   England.   The ROTHER in Sussex.
             The ROTHER, joins the Thames at
             Rotherhithe.

5.   *With the ending el.*
   Germany.   RAOTULA, 8th cent., now the RÖTEL.

Allied to the last word is the Eng. *race*,
and the many cognate words in the Indo-
European languages which have the sense
of rapid motion, as Welsh *rhysu,* &c.

1. *Scotland.*   The RASAY.   Rosshire.
   Ireland.   The ROSS.
   Germany.   The RISS.   Wurtemburg.
   Switzerland.   The REUSS.   Joins the Aar.
   Russia.   The RASA.
   Spain.   The RIAZA.
   Asia Min.   RHESUS of Homer not identified.
   India.   RASA, the Sanscrit name of a
             river not identified.
2.   *With the ending el.*
   Germany.   The ROSSL(AU).   Joins the Elbe.
3.   *With the ending et.*
   Germany.   The REZAT.   Joins the Rednitz.

From the Gael, *garbh,* Welsh *garw,* violent, Armstrong derives the name of the
Garonne and other rivers.* The root seems

---

* It will be seen, however, that while admitting this root, I do not
place Garonne to it.

to be found in Sansc. *karv* or *karp*, Latin *carpo*, &c., implying violent action. The Lat. *carpo* is applied by the poets to denote rapid progress, as of a river, through a country. So likewise more metaphorically to the manner in which a bold and steep mountain rises from the valley. As also one of our own poets has said—

> Behind the valley topmost Gargarus
> Stands up and *takes* the morning—

Hence this root is found in the names of mountains as well as rivers—*e.g.*, the Carpathians (Carpātes), and the Isle of Carpăthus, which "consists for the most part of bare mountains, rising to a central height of 4,000 feet, with a steep and inaccessible coast."[*]

1. *Scotland.*	GARF water, a burn in Lanarkshire.	
	The GRYFFE. Renfrew.	
Germany.	The GRABOW. Pruss. Pom.	
Danub. Prov.	CARPIS, Herodotus, see p. 73.	
2.	*With the ending en.*	
Scotland.	The GIRVAN. Ayr.	
Italy.	The CARPINO. Joins the Tiber.	
	The GRAVINO. Naples.	

---

* Smith's Ancient Geography.

| 3. | | *With the ending el.* |
| | *Italy.* | CERBALUS* ant., now the CERVARO —here ? |

From the Sansc. *su*, to shoot forth, *sûs*, *sûtis*, rushing or darting, Gr. σουσις, cursus, I take to be the following. Among the derived words, the Gael. *sûth*, a billow, seems to be that which comes nearest to the sense required.

1.	*Switzerland.*	The SUSS.
	Denmark.	The SUUS(AA).
	Bohemia.	The SAZ(AWA). Joins the Moldau.
	Portugal.	The SOUZA.
	Siberia.	The SOS(VA), two rivers.
	India.	The SUT(OODRA), or Sutledge— here ?†
2.		*With the ending en.*
	France.	The SUZON.
	Russia.	The SOSNA, two rivers.

Probably to the above we may put a form *sest, sost,* found in the following.

1.	*Germany.*	The SOESTE. Oldenburg.
	Italy.	SESSITES ant., now the Sesia.
	Persia.	SOASTUS or SUASTUS ant.

---

* This river of Apulia, though small in summer, is exceedingly violent in winter.

† "In its upper part it is a raging torrent."

Johnston's Gazetteer.

2.   *With the ending er.*
Russia.   The SESTRA. Gov. Moskow.
Germany.   The SOSTER(BACH). Joins the Lippe.

To the above root I also place the following, corresponding more distinctly with Old
High German *schuzzen,* Ang.-Sax. *sceotan,*
Eng. *shoot,* Obs. Gael. and Ir. *sciot,* dart,
arrow.*

1.   *With the ending en.*
Germany.   SCUZNA, 8th cent., now the SCHUS-
             SEN.
             SCUZEN ant., now the SCHOZACH.

2.   *With the ending en.*
Germany.   SCUTARA, 10th cent., now the SCHUT-
             TER, two rivers.
             SCUNTRA, 8th cent., now the SCHON-
             DRA and the SCHUNTER.

From the Germ. *jagen,* to hunt, to drive
or ride fast, Bender derives the name of the
Jaxt, in the sense of swiftness, suggesting
also a comparison with the ancient Jaxartes
of Asia. Förstemann considers both sugges-
tions doubtful, but the former seems to me
to be reasonable enough. The older sense

---

* The derivation of Mone, who makes *scuz* and *scut* altered forms of
srot or *srut,* is not to be entertained.

of *jagen* is found in the Sansc. *yug*, to dart forth, formed on the simple verb *ya*, to go. And appellatives are found in the Finnic words *jokk*, *jöggi*, a river. As for the Jaxartes, I am rather inclined to think that the more correct form would be Jazartes, and that it contains the word *jezer*, before referred to.

1.	*Russia.*	The JUG.	Joins the Dwina.
2.		*With the ending et.*	
	Italy.	JACTUS ant.	Affluent of the Po.
	Persia.	The JAGHATU.	
	Germany.	The JAHDE,* in Oldenburg.	
3.		*With the ending st.*	
	Germany.	JAGISTA ant., now the JAXT or JAGST.	

From the root *vip*, to move, p. 64, by the prefix *s*, is formed Old Norse *svipa*, Ang.-Sax. *swifan*, Eng. *sweep*, &c. In these the sense varies between going fast and going round, and the same may be the case in the following names.

---

* I am not sure that the Jahde of Oldenburg does not contain the more definite idea of a horse (Eng. *jade*, North. Eng. *yawd*). There are three rivers near together, the Haase, the Hunte, and the Jahde. It rather seems as if the popular fancy had got up the idea of a hunt, and named them as the Hare, the Hound, and the Horse.

France.   The SUIPPE. Joins the Aisne.
Germany.   SUEVUS, 2nd cent., now the Warnow,
             or, according to Zeuss, the Oder.
             SUAB(AHA), 8th cent., now the
             SCHWAB(ACH).

From the Obs. Gael. *sgiap, sgiob,* to move rapidly, Eng. *skip,* may be the following.

1. *England.*   The SHEAF, by Sheffield.
   Germany.   SCIFFA, 9th cent., now the SCHUPF.
   Asia Min. SCOPAS ant., now the Aladan.

2.   *With the ending en.*
   England.   The SKIPPON. Joins the Wyre.

In the Gael. *brais,* impetuous, related perhaps to Lat. *verso,* we may find the root of the following.

1. *Germany.*   The BIRSE. Prussia.
   Switzerland. The BIRSE. Cant. Berne.

2.   *With the ending en.*
   Ireland.   The BROSNA. Leinster.
   Transylvania. The BURZEN.   Joins the Aluta.
   Pruss. Pol.   The PROSNA.

3.   *With the ending el.*
   France.   The BRESLE. Enters the English
                 Channel.

4.   *With the ending ent.*
   Germany.   The PERSANTE. Pruss. Pom.

From the Sansc. *rab* or *rav*, to dart forth, whence (in a somewhat changed sense) Eng. *rave*, French *ravir*, Lat. *rabidus*, &c. The original meaning of a ravine was a great flood, or as Cotgrave expresses it—"A ravine or inundation of water, which overwhelmeth all things that come in its way."

1. *Ireland.*	The ROBE.	Connaught.
India.	The RAVEE or Iraotee—here ?	
2.	*With the ending en.*	
England.	Various small streams called RAVEN, RAVENBECK, &c.	
France.	The ROUBION, affluent of the Rhone —here ?	

From the Sansc. *math*, to move, are derived, as I take it, Old High German *muot*, Mod. Germ. *muth*, Ang.-Sax. *môd*, courage or spirit, Welsh *mwyth*, swift, &c., to which I place the following.

1. *Switzerland.*	The MUOTTA.	Cant. Schwytz.
2.	*Compounded with vey, stream or river.*	
Wales.	The MUTHVEY.	Three rivers.

The Sansc. *sphar, sphurj*, to burst forth, shews the root of a number of words such

as *spark, spring, spirt, spruce, spry,* in which
the sense of briskness or liveliness is more
or less contained. But the Sansc. *sphar* or
spar must be traced back to a simpler form
spa or *spe,* as found in *spew,* to vomit, and
in the word *spa,* now confined to medicinal
springs.

1. *Scotland.* The SPEY. Elgin.

2. *With the ending en.*
   Scotland. The SPEAN.

3. *With the ending er.*
   Scotland. The SPEAR.
   Germany. SPIRA, 8th cent., now the SPEIER.
   The SPREE. Joins the Havel.

Derived forms from the above root are
also the following, which correspond more
closely with Germ. *sprütsen,* Ang.-Sax. *spry-
tan,* Eng. *spirt,* Ital. *sprizzare.* And I think
that most of these names are probably German.

England. The SPRINT, a small stream in West-
moreland.
Germany. SPRAZAH, 9th cent., some stream in
Lower Austria.
The SPROTTA in Silesia.

SPRENZALA, 8th cent., now the SPREN-
ZEL.

SPURCHINE(BACH),* 9th cent., now
the SPIRCKEL(BACH).

Eu. Turkey. The SPRESSA. Joins the Bosna.

In the preceding chapter I have treated
of the root *al, el, il,* to go, and various of its
derivations. There is another, *alac, alc, ilc,*
which, as it seems most probably either to
have the meaning of swiftness, as in the Lat.
alacer, or of tortuousness, as in the Greek
ἑλικος, I include in this place.

1. *Russia.*   The ILEK. Joins the Ural.
   Sicily.   HALYCUS ant., now the Platani.
   Asia Minor. ALCES ant. Bithynia.

2.   *Compounded with may, main, river.*
   Siberia.   The OLEKMA. Joins the Lena.
   Germany.   ALKMANA, 8th century, now the
                  Altmühl.
   Greece.   HALIACMON ant., now the Vistritsa.

From the Welsh *tarddu,* to burst forth,
we may take the following. There does not
seem any connection between this and the

---

* Förstemann derives this, along with some other local names, from
Old High Germ. *spurcha,* the juniper-tree. But I think that the stream at
least is to be explained better from the Sansc. *sphurj,* to burst forth, Lat.
spargo.

root of *dart* (jaculum) ; the latter from the first signifies penetration, and in river-names comes before us in the oblique sense of clearness or transparency.

1. *Scotland.* The TARTH. Lanarkshire.
  Libya. DARĂDUS ant., now the Rio di Ouro.
  Armenia. DARADAX* ant. (Xenophon).
2. *With the ending er.*
  France. The TARDOIRE. Dep. Charente.
  Aust. Italy. The TARTARO.
3. *With the ending es.*
  Spain. TARTESSUS ant., now the Guadalquiver.

With the Sansc. *till*, to move, to agitate, we may probably connect the Gael. *dile* and *tuil*, Welsh *diluw, dylif, dylwch*, a flood, deluge, as also Ang.-Sax. *dilgian*, German *tilgen*, to overthrow, destroy, &c. The Ang.-Sax. *dêlan*, Germ. *thielen*, to divide, in the sense of boundary, may however intermix in these names.

1. *England.* The TILL. Northumberland.
  Ireland. The DEEL. Limerick.
  Germany. The DILL. Nassau.

---

* The ending *x* I take to be a Græcism for *t*.

> *Belgium.*   THILIA, 9th cent., now the DYLE in
>   Bravant.
> *Switzerland.* The THIELE.
> 2.   *With the ending en.*
> *Germany.*   The TOLLEN.   Mecklenburg-Schwe-
>   rin.
> 3.   *With the ending er.*
> *Scotland.*   The DILLAR burn.   Lesmahagow.
> 4.   *With the ending es.*
> *Germany.*   The TILSE, by Tilsit.

With the two Welsh forms *dylif* and *dylwch*, deluge, we may perhaps connect the following, though for the former the Ang.-Sax. *delfan*, to dig, *delf*, a ditch, may also be suitable.

> *Germany.*   DELV(UNDA), 9th century, now the
>   DELVEN(AU).
>   DELCHANA, 11th century, now the
>   DALCKE.

From the Gael. and Ir. *taosg*, to pour, *tias*, tide, flood, may be the following. Perhaps the special sense of cataract may come in, at least in some cases, as two of the under-noted rivers, the Tees and the Tosa, are noted for their falls.

> 1. *England.*   The TEES.   Durham.

Switzerland. The Töss. Cant. Zurich.
Piedmont. The Tosa.
Russia. The Tescha. Joins the Oka.
Hungary. Tysia ant., now the Theiss.
Greece. Tiasa ant. Laconia.
India. The Touse—here ?

2. *With the ending en.*
Switzerland. The Tessin or Ticino.
Germany. The Desna. Joins the Dnieper.
France. The Tacon. Dep. Jura.

3. *With the ending el.*
Germany. Tussale (*Genitive*), 11th cent., now the Dussel by Düsseldorf.

4. *With the ending st.**
England. The Test. Hants.
Germany. The Dista. Prussia.
India. The Teesta—here ?

From the Sansc. *gad* or *gand*, Ang.-Sax. *geôtan*, Suio-Goth. *gjuta*, Danish *gyde*, Old Norse *giosa*, Old High Ger. *giezen*, Obs. Gael. *guis*, all having the meaning of Eng. "gush," we get the following. The Gotha or Gœta of Sweden may probably derive its name from the well-known fall which it makes at

---

* In these names we may perhaps think of the Bohem. *dest*, rain. The Teesta is much swollen in the rainy season, but perhaps not more so than most of the other rivers of Hindostan. In Hamilton's East Indian Gazetteer, it is explained as "*tishta*, standing still,"—a derivation which seems hardly to agree with the subsequent description of its "quick stream."

Trolhætta.   So also the Gaddada of Hindostan is noted for its falls ; and the Giess-
bach is of European celebrity.   But in some
of the other names the sense may not extend
beyond that of wandering, as we find it in
Eng. *gad*, which I take to be also from this
root.   Or that of stream, as in Old High
Germ. *giozo*, Gael. and Ir. *gaisidh*, rivulus.

1. *England.*   The GADE.   Herts.
   Scotland.   GADA ant.,* now the JED byJedburgh.
   Germany.   The GOSE.   Joins the Ocker.
             GEIS(AHA), 8th cent., now the GEISA.
             The GANDE, Brunswick—here, or to
             can, cand, pure.?
   Switzerland. The GIESS(BACH).   Lake of Brienz.
   Spain.   The GATA.   Joins the Alagon.
   Sweden.   The GOTHA or GŒTA.
             The GIDEA, enters the G. of Bothnia.
   Asia.   GYNDES (*Herodotus*), perhaps the
             Diala—here ?

2.   *With the ending en.*
   Asia Minor. CYDNUS ant., now the Tersoos Chai.

3.   *With the ending er.*
   Persia.   The GADER.
   Sardinia.   CÆDRIUS ant., now the Fiume dei
             Orosei.

---

* Hence Baxter derives the name of the Gadeni—" Quid enim Gadeni
nisi ad Gadam amnem geniti ?"

4.   *With the ending el.*
  Germany.   Gisil(AHA), 8th cent., now the GIESEL
              —here ?

5.   *With the ending ed.*
  India.   The GADDADA.

6.   *Compounded with main, stream.*
  Switzerland. The GADMEN.

From the Sansc. *arb* or *arv*, to ravage or destroy, cognate with Lat. *orbo*, &c., may be the following. To the very marked characteristic of the Arve in Savoy I have referred at p. 6. But there is a word of precisely opposite meaning, the Celt. *arab*, Welsh *araf*, gentle, which is very liable to intermix.

1. *France.*   The ARVE and the ERVE.
   Germany.   ORB(AHA), 11th cent., now the ORB.
   Sardinia.   The ARVE and the ORBE.
   Hungary.   The ARVA. Joins the Waag.
   Spain.   The ARVA, three rivers, tributaries
                 to the Ebro.

2.   *With the ending en.*
   Scotland.   The IRVINE. Co. Ayr.
   France.   ARVENNA ant., now the ORVANNE.

3.   *With the ending el.*
   Germany.   ARBALO, 1st cent., now the ERPE.

4.   *With the ending es.*
   Asia Minor. HARPĂSUS ant., now the HARPA.

In the Sansc. *cal*, to move, and the derivatives Sansc. *calas*, Gr. κελης, Obs. Gael.
callaidh, Latin *celer*, all having the same
meaning—the sense of rapidity seems sufficiently marked to include them in this
chapter.

1. *Scotland.*   The GALA.   Roxburgh.
   Sicily.   GELA ant.*
   Illyria.   The GAIL.
   Greece.   CALLAS ant., in Eubœa.
   As.-Turkey. The CHALUS of Xenophon, now the
                 Koweik.
2.   *With the ending en.*
   Ireland.   The CALLAN.   Armagh.
3.   *With the ending er = Lat. celer ?*
   Italy.   CALOR ant., now the CALORE.
4.   *With the ending es = Sansc. calas, &c. !*
   Germany.   CHALUSUS, 2nd cent., supposed to
                 be the Trave.
                 The KELS, in Bavaria.
   India.   The CAILAS.

I am inclined to bring in here, as a derivative form of *cal*, and perhaps corresponding
with the Obs. Gael. *callaidh*, celer, the forms

---

* The Gela is at times a very violent stream, as the following description of Ovid bears witness.
         "Et te vorticibus non adeunde Gela."
                              Fasti. 4, 470.

caled, calt, gelt.   That the Germ. *kalt*, Eng.
cold, may intermix, is very probable, but I
do not think that all the English rivers at
any rate can be placed to it.   There is more
to be said for it in the case of the Caldew
than of the others, for one of the two streams
that form it is called the Cald-beck (*i.e.*, cold
brook), and it seems natural that the whole
river should then assume the name of Caldew (cold river).   Yet there may be nothing
more in it than that the Saxons or Danes
who succeeded to the name, adopted it in
their own sense, and *conformed* to it.   It is to
be observed that although the form Caldew
corresponds with the Germ. Chaldhowa, yet
that the local pronunciation is invariably
Cauda (= Calda), corresponding with the
Scandinavian form.   Upon the whole however, there is much doubt about this group ;
the form *gelt* Förstemann refers, as I myself
had previously done, to Old Norse *gelta*, in
the sense of resonare.   In the following
names I take the Kalit(va) of Russia, and

the Celydnus and Celadon of Greece to approach the nearest to the original form.

1.	*England.*	`The GELT. Cumberland.
		The CHELT by Cheltenham—here ?
		The CALD(EW). Cumberland.
	Germany.	The CALD(HOWA), (*Adam Brem.*), now seems to be called the Aue.
	Russia.	The KALIT(VA). Joins the Donetz.
2.		*With the ending en.*
	Germany.	GELTEN(AHA), 11th cent, now the GELTN(ACH).
	Greece.	CELYDNUS ant. Epirus.
		CELADON ant. Elis.
3.		*With the ending er.*
	England.	The CALDER. Three rivers.
	Scotland.	The CALDER. Joins the Clyde.
	Belgium.	GALTHERA, 9th cent.

I am also inclined to bring in, as another derivative form of *cal,* the word *calip, calb, kelp.* The only appellatives I find for it are the word *kelp,* sea-weed, and the Scottish *kelpie,* a water-spirit, wherein, as in other words of the same sort, may perhaps lie a word for water. However, this can be considered as nothing more than a conjecture.

1. *Germany.*   KALB(AHA), 8th cent., now the Kohlb(ach).

                  The KULPA. Aust. Croatia.

Hungary.   COLAPIS ant., affluent of the Drave.

Spain.   The CHELVA. Prov. Valentia.

Portugal.   CALLÍPUS ant., now the Sadao.

Asia Minor. CALBIS ant. Caria.

                  CALPAS ant. Bithynia.

2.   *With the ending en.*

Scotland.   The KELVIN. Stirling.

The Sansc. *car,* to move, Lat. *curro,* like some other words of the same sort, branches out into two different meanings—that of going fast, and that of going round. Hence the river-names from this root have in some cases the sense of rapidity, and in others of tortuousness; and these two senses are somewhat at variance with each other, because tortuousness is more generally connected with slowness. Separating the two meanings as well as I can, I bring in the following here.

1. *Scotland.*   The GARRY. Perthshire.

                  The YARROW. Selkirkshire.

2.   *With the ending en.*

England.   GARRHUENUS ant., now the YARE.

P

France.	GARUMNA or GARUNNA ant. The GARONNE.
	The GIRON. Joins the Garonne.
Greece.	GERANIUS ant., and GERON ant., two rivers of Elis, according to Strabo.

3.   *With the ending es = Sansc. caras, swift, Lat. cursus, &c.*

France.	The GERS. Joins the Garonne.
	CHARES ant., now the CHIERS.
Germany.	The KERSCH. Joins the Neckar.
Italy.	The GARZA, by Brescia.
Hungary.	GERĂSUS ant., now the KOROS.
Asia Minor.	The CARESUS of Homer in the plain of Troy.
Syria.	CERSUS ant., now the Merkez.

There appear to be several words in which the sense of violence or rapidity is brought out by the preposition *pra, pro, fro,* in composition with a verb. Thus the Welsh *ffre-uo,* to gush, whence *ffrau,* a torrent, seems to correspond with the Sansc. *pra-i,* Lat. *præ-eo,* &c. Or perhaps we should take a verb with a stronger sense, say *yu,* to gush, and presume a Sansc. *pra-yu* = Welsh *ffre-uo.* In the Albanian πρό, a torrent, corresponding with Welsh *ffrau,* there seems, however, no trace of a verb.

1. *Wales.*   The FRAW, by Aberfraw.
2.   *With the ending en.*
   Scotland.   The FROON.   Falls into L. Lomond.
   Russia.   The PRONIA.

The Welsh *ffrydio,* to stream, to gush, appears to be formed similarly from the preposition *fra,* joined with the verb *eddu,* to press on, to go, corresponding with Sansc. *it,* Latin *ito,* &c.   Hence it would correspond with a Sansc. *pra-it,* Lat. *præ-ito,* &c.   From the verb comes the appellative *ffrwd,* a torrent, corresponding with the Bohem. *praud,* of the same meaning.

Scotland.   The FORTH.   Co. Stirling.
Danub. Prov. PORATA (Herodotus).   The PRUTH.
Russia.   The PORT(VA).   Gov. Kaluga.

I also bring in here, as much suggestively as determinately, the following.

Sansc. pra-pat, Lat. præ-peto, &c., to rush forth.
Russ. Pol.   The PRIPET.   Joins the Dneiper.
Bulgaria.   The PRAVADI.   Falls into the Black Sea.
Sansc. pra-cal, to rush forth, pra and cal, p. 112.
Prussia.   The PREGEL.   Enters the Frische-Haff.
Sansc. pra-li, Lat. pro-luo, &c., to overflow.
India.   The PURALLY.

According to the opinion of Zeuss and Gluck, the DANUBE, (ant. Danubius and Danuvius, Mod. Germ. Donau,) would come in here. These writers derive it from Gael. *dan*, Ir. *dana*, fortis, audax, in reference to its strong and impetuous current. This is no doubt the most striking characteristic of the river, but it might also not inappropriately be placed to the root *tan*, to extend, whence the names of some other large rivers. Gluck considers the ending *vius* to be simply derivative, and suggests that the Germans, with a natural striving after a meaning, altered this derivative ending into their word *ava, aha, ach,* or *au,* signifying river. Though Gluck is a writer for whose opinion I have great respect, and though this is the principle for which I myself have been all along contending, yet I am rather inclined to think that in Danuvius, as in Conovius (the Conway), there is contained a definite appellative, qualified by a prefixed adjective : this seems to me to be brought out more clearly in the Medway, and in the names connected with it.

The word Ister, which, according to Zeuss, is the Thracian name of the Danube, I have elsewhere referred to the Armorican *ster*, a river. Not that I mean to infer therefrom that the name is Celtic, because *ster* is only a particular form of an Indo-European word *sur*. If we refer the prefix *is* to the Old Norse *isia*, proruere, then Ister would have the same meaning as that given above to Danubius. But the derivation of Mone, who explains it by *y*, the Welsh definite article, and *ster*, a river, making Ister = " The river," I hold with Gluck to be—like other derivations proceeding on the same principle— opposed to all sound philology.

Among the rivers noted for the slowness of their course, the most conspicuous is the Arar or Saone. Cæsar (*de Bell. Gall.*) describes it as flowing " with such incredible gentleness that the eye can scarcely judge which way it is going." Seneca adopts it as

a type of indecision—"the Arar in doubt
which way to flow." Eumenius multiplies
his epithets—"segnis et cunctabundus amnis,
tardusque." The name Sauconna, Sagonna,
Saonna, Saone, does not appear before the
4th cent., yet there does not seem any reason
to doubt that it is as old as the other. Zeuss
(*Die Deutschen*) and the Editor of " Smith's
Ancient Geography" take this as the true
Gallic name. And though Armstrong explains both the Arar and the Saone from the
Celtic—referring the former to the Obs.
Gael. *ar*, slow, and the latter to Gael. *sogh*,
tranquil or placid, in which he may probably
be correct, yet it by no means follows that
the name of the Arar is Celtic, for *ar* is an
ancient root of the Indo-European speech.
To the same root as the Saone I also put
the Seine (Sequăna), and the Segre (Sicŏris),
comparing them with Lat. *seg-nis*. The
former of these rivers is navigable for 350
miles out of 414, and the latter is noted in
Lucian as "stagnantem Sicorim." Some

other rivers, in which the characteristic is less distinct, I also venture to place here, separating this root as well as I can from another p. 58.

1. *Germany.*	SIGA, 10th cent. The SIEG.	
Russia.	The SOJA. Joins the Dneiper.	
2.	*With the ending en.*	
France.	SAUCONNA ant. The SAÔNE.	
	SEQUANA ant. The SEINE.	
	The SEUGNE. Dep. Charente-Inf.	
Russia.	The SUCHONA. Joins the Dwina.	
3.	*With the ending er.*	
Spain.	SICORIS ant. The SEGRE.	
	The SEGURA. Enters the Med. Sea.	

Perhaps allied in its root to the last is the Gael. *saimh*, quiet, tranquil, to which I put the following.

1. *Belgium.*	The SEMOY.	
Russia.	The SEM or SEIM. Joins the Desna.	
	SAIMA, a lake in Finland.	
Asia Minor.	The SIMOIS of Homer—here ?	
2.	*With the ending en.*	
Switzerland.	The SIMMEN, in the Simmen-Thal.	
3.	*With the ending er.*	
France.	SAMARA ant., now the SOMME.	
	The SAMBRE, ant. Sabis.	
Germany.	The SIMMER. Joins the Nahe.	
Russia.	The SAMARA. Two rivers.	

4. *With the ending et.*
Germany. SEMITA, 8th cent. The SEMPT.

In the Gael. *ar*, slow, (whence the Arar, p. 118,) is to be found, as I take it, the root of the Welsh *araf*, mild, gentle. From this Zeuss (*Gramm. Celt.*), derives the name of the Arrăbo, now the Raab. This root is liable to mix with another, *arv*, p. 109, of precisely opposite meaning.

Hungary. ARRABO ant., now the Raab.
India. ARABIS ant., now the Purally.
Ireland. The AROB(EG),* Co. Cork—here ?

I bring in here the word *aram* or *arm*, which, both in the names of rivers, and in the ancient names of men, as the German hero Arminius, needs explanation. The authority of Dr. Donaldson may probably have been the cause of the reproduction, even in some of the latest English works, of the mistake of confounding the name Armin, Ermin, or Irmin, with the word *hermann*, warrior, (from *her*, army, *mann*,

---

* This ending may be the same as the Scotch *eck* or *ick*, p. 25.

homo). That it is not so is shown by its appearance in the ancient names of women, as Ermina, Hermena, and Irmina,* (daughter of Dagobert the 2nd). And by the manner in which it forms compounds, as Armenfred, Irminric, Irminger,† Ermingaud, Irminher, &c. For we may take it as a certain rule that no word, itself a compound, forms other compounds in ancient names. Indeed, the last of the five names, Irminher, (which is found as early as the 7th cent.), is formed from the word *her*, army, so that, according to the above theory, it would be Her-mann-her. The fact then, as I take it, is that, both in the names of rivers and of men, the root is simply *arm* or *irm*, and *armin* or *irmin* an extended form, like those found all throughout these pages. As to its etymology, the word *aram*, *arm*, in the Teutonic dialects signifying poor or weak, is in

---

* Förstemann, Altdeutsches Namenbuch. (Vol. 1. Personennamen).

† The names ARMINE and ARMINGER, (of which IREMONGER may be a corruption), occur in Lower's Patronymica Britannica. And ARMINGAUD is one of the many names of German or Frankish origin still found in France.

itself unsuitable, but I think that the original meaning may perhaps rather have been mild or gentle. The root seems to be found in the Gael. *ar*, slow ; and *aram* may be a corresponding word to the Welsh *araf.* Baxter, who, though his general system of river-names I hold to be fallacious, was, for his time, no contemptible etymologist, suggests something of the sort.

1. *England.*   The ARME.   Devon.
   Russia.   The URJUM(KA)—here ?
2.   *With the ending en.*
   Italy.   ARIMINUS ant., now the Marecchia.
                 The ARMINE.
3.   *With the ending es.*
   Germany.   ARMISIA ant., now the ERMS.

In this place I am inclined to bring in the Medway, and some other names connected with it. Among the various derivations which have been suggested for this name, that of Grimm deserves the first place, though I much fear that it is too poetical to be true. He observes, (*Gesch. d. Deutsch. sprach.*), comparing it with another name—

"In Carl's campaign, A.D. 779, there is a place mentioned in the vicinity of the Weser, called Medofulli, Midufulli ; *medoful* means poculum mulsi, (*Hel.* 62, 10) ; it appears to have been a river, which at present bears some other name. Of just a similar meaning is the name of the river Medway flowing through the county of Kent into the Thames —*i. e.*, Ang.-Sax. Meadovaege, Medevaege Medvaege (*Cod. Dip.*), from *vaege*, Old Sax. *wêgi*, Old Norse *veig*, poculum. . . I suggest here a mythological reference : as the rivers of the Greeks and Romans streamed from the horn or the urn of the river-god, so may also the rivers and brooks of our ancestors, in a similar mythic fashion, have sprung from the over-turned mead-cup."

It is a pity to disturb so poetical a theory, coming too as it does from the highest authority, but I much fear that on a comparison of this name with all its related forms, it can hardly be substantiated. For the word does not stand alone—the prefix *med* is

found in several names in which the second
part can hardly be taken to mean poculum,
and the ending *way* is found in several names
of which the former part cannot mean mulsum.   In any case, it seems to me that a
Saxon derivation can hardly be sustained.
For Medoăcus, (= Medwacus), occurs as the
ancient name of a river in Venetia—this appears to be precisely the same name as that
of the Medwag or Medway—and in Venetia
we can account for a Celtic element, but not
for a German.   In Nennius the name stands
as Meguaid or Megwed ; and comparing this
with a river called the Medvied(itza) or
Medviet(za) in Russia, it would seem rather
probable that the form is not altogether false,
but that only it should be Medwed instead
of Megwed.   In that case it would probably
be only another form of Medweg, for *d* and
g sometimes interchange in the Celtic dialects, as in the Gaelic *uidh* and *uigh*, via, a
word which indeed I take to be related to
the one in question.   Again, in the Med-

uāna of France and the English Medwin,
we have a third form of ending, *wân* or
win. And this may probably only be one
of those extended forms in *n* so common
in the Celtic languages.* So that the
endings *way, wân, wied,* in Medway, Meduāna, Medvied(itza), may be slightly differ-
ing forms of a common appellative (p. p. 62,
63), qualified by the prefix *med,* which we
have next to consider. In Gibson's "Etymological Geography" *med* is explained as
medius — Medway = medium flumen — the
river flowing through the middle of the
county of Kent—and this I think is the
general acceptation. In the case of the Medina, (ant. Mede), which divides the Isle of
Wight into two equal parts, I should readily
accept such a derivation, but in the case of
the Medway it seems to me a feature scarcely
sufficiently obvious to give the name. And
I should on the whole prefer a derivation

---

* E. G. Welsh *lli, llion,* stream, *llif, llifon,* flood, *srann, srannan,*
humming, &c.

from the same root as mead, mulsum, viz.,
Sansc. *mid*, to soften, Lat. *mitis*, Gael. *meath*,
soft, mild—finding in Old Norse *mida*, to
move slowly or softly, the word most nearly
approximating to the sense, and thus deriving the name of the Medway from its gentle
flow.

Nevertheless it must be observed that as
well as the supposed river Medofulli referred
to as above by Grimm, we find in a charter
of the 10th cent., a river called Medemelacha,
which seems evidently to contain the Gael.
mealach, sweet, and to mean "sweet as
mead." This river is near Medemblik on the
Zuyder-zee, and I suppose that the name of
the place is corrupted from it.

The following names I place here, though
with uncertainty in the case of some of them.

1. *France.*   The MIDOU.   Dep. Landes.
   Persia.   MEDUS ant., now the Pulwan.
2.   *With the ending en.*
   Russia.   The MEDIN(KA).   Gov. Kaluga.
3. *Compounded with way, wân, wied, see above.*
   England.   The MEDWAY.   Kent.

England. The MEDWIN.
France. MEDUĀNA ant., now the Mayenne.
Italy. MEDOÏCUS ant., now the Brenta.
Russia. The MEDVIED(ITZA).

4. *Compounded with ma, river, p.* 60.
Germany? METEMA, in a charter of the 11th cent.

I think, upon the whole, that the general meaning of the root *lam, lem, lim,* is smoothness. Though the root-meaning seems rather that of clamminess or adhesiveness, as found in Sansc. *limpas,* Gr. λιπος, Lat. *limus,* Old Sax. *lêmo,* Mod. Germ. *lehm,* Eng. *lime,* &c.*
In the Gr. λιμνη, lake, the sense becomes that of smooth or standing water : this, as I take it, is in effect the word found in the Lake Leman, Loch Lomond, &c. Though the word most immediately concerned is the Gaelic *liobh, liomh,* Welsh *llyfnu,* to smooth ; and the Loch Lomond, (properly Lomon), was also formerly called, as the river which issues from it is still, Leven, being just another form of the same word—*v* and *m* inter-

---

* Hence perhaps Lemanaghan, a parish of Leinster, which consists chiefly of bog.

changing as elsewhere noticed. Hence the Welsh *llifo*, to pour, p. 46, might be apt to intermix in the following. The Lat. *lambo*, the primitive meaning of which is to lick, is applied to the gentle washing of a river against its banks—" Quæ loca lambit Hydaspes," —*Horace*. Dugdale observes that " at this day divers of those artificial rivers in Cambridgeshire, anciently cut to drain the fens, bear the name of Leam, being all muddy channels through which the water hath a dull or slow passage." In the following names the sense may be sometimes then that of muddiness, though in general, as I take it, that of sluggishness.

1. *England.*	The LEAM by Leamington.
	The LYME. Dorsetshire.
Germany.	LAMMA, 11th cent. The LAMME.
	LAIM(AHA), 8th cent. Not identified.
	LEMPHIA, 8th cent. The LEMPE.
Russia.	The LAMA. Joins the Volga.
	The LAM(ov). Gov. Penza.
Italy.	The LIMA. Joins the Serchio.
Spain.	LIMÆA ant., now the LIMA.

Asia Minor. LAMUS ant., in Cilicia.

2.   *With the ending en.*

England.   The LEMAN.   Devonshire.
                 The LIMEN in Kent.   (Limeneâ
                     Cod. Dip.)
Scotland.   Loch LOMOND, formerly also called
                     LEVEN.
Switzerland. Lake LEMAN, or the Lake of Geneva,
                     (ant. LEMANNUS.)
Italy.   The LAMONE in Tuscany.

3.   *With the ending er.*
Germany.   LAMER, 11th cent.   The LAMMER.
Italy.   The LAMBRO.
Asia Minor. LIMYRUS ant., in Lycia.

4.   *With the ending et.*
Switzerland. The LIMMAT.   Cant. Zurich.

From the above form *lam, lem, lim,* I take
to be formed by metathesis *alm, elm, ilm.*
And the lake Ilmen in Russia I take to be
in effect the same word as the lake Leman
in Switzerland.   In the name of another lake
in Russia, the Karduanskoi-ilmen, it seems
to occur as an appellative.   A certain amount
of doubt is imported by the coincidence of
two names in which we find a sacred character—the river Almo, which was sacred to

R

Cybele, and a sacred fountain Olmius mentioned in Hesiod. The coincidence, however,
may be only accidental.

1.	*England.*	The ALME. Devonshire.
		The HELME. Sussex.
		ALUM Bay in the Isle of Wight ?
	Germany.	ILMA, 8th cent. The ILM, two rivers.
		The HELME in Prussia.
	Holland.	The ALM in Brabant.
	Norway.	The ALMA.
	Spain.	The ALHAMA. Prov. Navarra.
	Italy.	The ALMO near Rome.
	Russia.	The ALMA in the Crimea.
	Siberia.	The ILLIM.
	Greece.	OLMEIUS ant. Bœotia.
2.		*With the ending en.*
	Gemany.	The ILMEN(AU). Joins the Elbe.
	Russia.	ILMEN. Lake.
3.		*With the ending el.*
	Holland.	The ALMELO. Prov. Overijssel.

Perhaps from the Gael. *foil,* slow, gentle,
we may get the following.

1.	*England.*	The FAL by Falmouth.
	Ireland.	The FOIL(AGH). Cork.
		The FEALE. Munster.
2.		*With the ending en.*
	Scotland.	The FILLAN. Perthshire.
3.		*With the ending es.*
	Germany.	FILISA, 8th cent. The FILS and the VILS.

In the third division of this chapter I put the names in which the sense of spreading seems to be found. This sense may have three different acceptations—first, that, generally, of a wide river—secondly, that of a river relatively broad and shallow—thirdly, that of a river forming an estuary at its mouth.

I bring in here the Padus or Po, which, by Metrodorus Scepsius, a Greek author quoted by Pliny, has been derived from the pine-trees, "called in the Gallic tongue *padi*," of which there were a number about its source. A derivation like this jars with common sense, for it is unreasonable to suppose that the Gauls, coming upon this fine river, gave it no name until they had tracked it up to its source, and there made the not very notable discovery that it was surrounded by pine-trees. Much more probable is it that they came first upon its mouth, and much more striking would be the appearance that would be presented to them.

For, as Niebuhr observes, "the basin of the Po, and of the rivers emptying themselves into it was originally a vast bay of the sea," which by gradual embanking was confined within its present channels. As then the mouth of the Padus was a vast estuary, so in the Gael. *badh*, a bay or estuary, I find the explanation of the name. The root, I apprehend, is Sansc. *pat*, Lat. *pateo*, *pando*, &c., to spread, and hence, I take it, the name Bander, of several small bays on the S.W. coast of Asia, of Bantry Bay in Ireland, and of Boderia, the name given by Ptolemy to the Firth of Forth.

1. *Italy.* PADUS ant. The Po.
   Germany. BADA, 9th cent., now the BODE.
2. *With the ending en.*
   Ireland. The BANDON. Co. Cork. (Forms a considerable estuary).
   Italy. PANTANUS ant., now the Lake of Lesina, a salt lagoon on the Adriatic.
3. *With the ending er.*
   Germany. PATRA, 9th cent., now the PADER.
4. *With the ending es.*
   Hungary. PATHISSUS ant., now the TEMES.*

* The names Pathissus and Temes I take to have the same meaning. I know no reason for supposing that the one name is less ancient than the other.

In the Sansc. *parth*, to spread or extend, we may perhaps find the origin of the following. Can the name of the Parthians be hence derived, in reference to their well-known mode of fighting ?

1. *Germany.*   The PARDE.   Joins the Elster.
   The BORD, in Moravia—here ?
2.   *With the ending en.*
   Asia Minor. PARTHENIUS ant.—here ?*

In the sense of " that which spreads" I am inclined to bring in the root *ta, tav, tan, tam.* While in the Gaelic we find *tain,* and the Obs. *ta,* water, *taif,* sea—in the Welsh we have the verbs *taenu* and *tafu,* to expand or spread. The latter, I think, must contain the root-meaning; and the appellatives must rather signify water of a spreading character. In this sense we find the words *to, tû, tau,* in the Hungarian dialects signifying a lake. The Sansc. has *tan,* to extend, but we must presume a simpler form *ta,* corresponding with the above Obs. Gael. word for water.

---

* The derivation of Strabo, from *parthenos,* virgin, in reference to the flowers on its banks, seems rather far-fetched.

Mone explains *tab*, as in Tabuda (the Scheldt), as "a broad river, especially one with a broad mouth." This sense no doubt obtains in many of the names of this group, for, as well as the Scheldt; the Tay, Taw, Teign, and Tamar, all have this character in a more or less notable degree. In other cases the sense may be that of comparative broadness—thus the Timavus, though little more than a mile long, is 50 yards broad close to its source. So the characteristic of the Dane, as noticed by the county topographers, is that it is "broad and shallow." And the feature which strikes the topographer is of course that which would naturally give the name. There are, however, some other roots which might intermix, as Sansc. *tan*, resonare, Lat. *tono*, Germ. *tönen*, &c. Also Gael. and Ir. *taam*, to pour; Gael. and Ir. *tom*, to bathe, Welsh and Ir. *ton*, unda.

The form Ta, Tab, Tav.

1. *England.*   The TAVY and the TAW. Devon.
     DEVA ant., the DEE—here ?

Scotland.	TAVUS ant. The TAY.
	The DEE, two rivers—here ?
Wales.	The TAW, the TIVY, and the TAVE.
Ireland.	The TAY. Waterford.
	Loch TA in Wexford.
France.	The DIVE, Dep. Vienne—here ?
Germany.	The THAYA in Moravia.
Spain.	The DEVA by Placentia—here ?

2.   *With the ending d or t.*

Scotland.	The TEVIOT in Roxburghshire—here ?
Holland.	TABUDA ant., now the Scheldt.
Siberia.	The TAVDA.
India.	The TAPTEE—here ?

The form Tan, Tam.

1. *England.*	The TEIGN and the TEANE.
	The DANE and the DEANE.
	The TAME, three rivers.
Scotland.	The TEMA. Selkirkshire.
	DANUS ant., now the DON.
France.	DANUS ant., now the AIN.
	The DAHME and the DÉAUME.
Norway.	The TANA.
Italy.	TIMAVUS ant., now the TIMAO.
Russia.	TANAIS ant., now the DON.
	The TIM and the TOM.
Greece.	TANUS ant., now the Luku.

2.   *With the ending er.*

England.	The TAMAR. Cornwall.
Belgium.	The DEMER.
Italy.	TANARUS ant., now the TANARO.
Spain.	TAMARIS ant., now the TAMBRE.

Syria.   TAMYRAS ant., (Strabo)—here ?

3.   *With the ending d.*
England.   TAMEDE (*Cod. Dip.*), now the TEME.
Mauretania. TAMUDA ant. (*Pliny.*)

4.   *With the ending es.*
England.   The THAMES. Tamesis (*Cæsar*), Tam-
                esa (*Tacitus*), Tamese, Temis (*Cod.
                Dip.*), Welsh Tain.
Hungary.   The TEMES, ant. Pathisus, (*see note
                p.* 132).

From the root *tan,* to extend, we may probably also derive the word *tang* found in
Hung. *tenger*, sea, Ostiakic (an Ugric dialect
of the Finnic class) *tangat*, river, and in the
Dan. *tang*, sea-weed, which probably contains a trace of an older sense.

1. *Holland.*   The DONGE in Brabant.
   Norway.   The TENGS.
2.   *With the ending er.*
   Germany. TONGERA, 10th cent., now the TAN-
                GER.
   Italy.   TANAGER ant., now the TANAGRO—
                here ?

CHAPTER VI.

CHARACTER OF COURSE.

In the inscription of Pul found at Nineveh, as deciphered in the Proceedings of the Asiatic Society, vol. 19, pt. 2, the Euphrates is called the Irat, which is conjectured by the translator to have been a local name. It seems to be from the Sansc. *irat* (= Latin *errans*, Eng. *errant*), from the verb *ir*, Lat. *erro*, to wander. The same word seems to be found in the Irati of Spain—perhaps also in the Orontes (= Irantes = Irates), of Syria. Possibly also in the Erid-anus or Po, though I am rather inclined to agree with Latham that the word contained therein is only *ridan.** Perhaps then the form Irt or Urt in

---

* That is, if it be the name of any real river falling into the Baltic, (the Rhodaune by Dantzic is suggested by some) ; but according to Heeren and Sir G. Lewis the Eridanus was a purely poetical stream, without any geographical position or character.—*See an article by Sir G. Lewis in Notes and Queries, July 3, 1S58.*

river-names may be a contracted form of *irat*, as we find it in the Germ. *irrthum*, a mistake.

1. *England.*   The IRT.   Cumberland.
                URTIUS ant., now the IRTHING.
   Belgium.   URTA, 9th cent., now the OURT.
                The ERENS.
   Spain.   The IRATI.   Prov. Navarra.
   Asia.   IRAT, a name of the Euphrates.
2.   *With the ending el.*
   Germany.   URTELLA, 9th cent., now the Sens-
                bach.

From the Sansc. *bhuj*, Goth. *bjugan*, Welsh *bwäu*, Gael. *bogh*, Eng. *bow*, &c., in the sense of tortuousness, we may take the following.

1. *England.*   The BOWE.   Shropshire.
   Scotland. The BOGIE.   Aberdeen.
   Russia.   The BUG.   Joins the Dnieper.
2.   *With the ending en.*
   Germany.   The BOGEN.   Joins the Danube.
3.   *With the ending et.*
   Scotland. The BUCKET.   Aberdeen.

From the Gael. and Welsh *cam*, to bend, Sansc. *kamp*, Gr. καμπω, are the following.

   England.   The CAM by Cambridge.
   Germany.   CAMBA, 8th cent.   The KAMP.
                The CHAM in Bavaria.

Switzerland. The KAM.
Norway. The KAM. Joins the Glommen.
Russia. The KAMA. Joins the Volga.
The KEMI. Two rivers.

The Sansc. root *car*, to move, branches out into two different meanings, that of rapidity and that of circuitousness, the former of which I have included in the previous chapter. In the latter sense we have the Gael. *car* or *char*, tortuous, the Ang.-Sax. *cêrran*, to turn or bend, &c., to which I place the following.

1. *England.* The CHAR. Dorsetshire.
The CHOR. Lancashire.
The KERR. Middlesex.
Scotland. COR(ABONA)* ant. The CARRON.
France. The CHER. Joins the Loire.
Greece. CHARES ant. Colchis.
Persia. CYRUS ant., now the KUR.
2. *With the ending en.*
England. CIRENUS ant. The CHURNE (Gloucestershire).
France. The CHARENTE.
3. *With the ending el.*
Greece. CORĂLIS ant. Bœotia.
CURALIUS ant. Thessaly.
Russia. The KOROL. Joins the Dnieper

---

* In this case the ending *en* is very clearly a contraction of *abon* or *avon*, river.

From the Old High Germ. *crumb*, Mod. German *krumm*, Danish *krumme*, Gael. and Welsh *crom*, curving or bending, we may take the following. The root seems to be found in the Sansc. *kram*, to move, to go, which, as in other similar cases, may also diverge into the meaning of rapidity.

1. *England.* The CRUMM(OCK), formerly CRUM-(BECK), which forms the lake of the same name.

Germany. CRUMB(AHA), 10th cent., now the GRUMB(ACH).

Russia. The KROMA. Gov. Orel.

2. *With the ending en.*

Germany. CHRUMBIN(BACH), 8th cent., now the KRUM(BACH).

3. *With the ending er.*

Italy. CREMERA ant. in Etruria.

4. *With the ending es.*

Germany. The KREMS. Joins the Danube.

Sicily. CREMISUS ant.

For the root *sid* we have the Welsh *sid*, winding, and the Anglo-Saxon *sid*, broad, spreading. The former is, I think, the sense contained in the following, though both words may be from the same root.

1. *England.* The SID. Devonshire.
2.   *With the ending en.*
  England. The SEATON. Cornwall.
3.   *With the ending rn, p. 34.*
  Switzerland. SITERUNA, 8th cent., now the SITTER
            or SITTERN.

Baxter's derivation of the Derwent from Welsh *derwyn,* to wind, appears to me the most suitable. That of Zeuss (taking the form Druentia), from *dru,* oak, seems insufficent; because the number of names, all in the same form, seem to indicate that the word contained must be something more than *dru.* That of Armstrong, from *dear,* great, *amhain,* river, is founded upon a careless hypothesis that the Derwent of Cumberland is the largest river in the North of England, which is not by any means the case.

England. The DERWENT. Four rivers.
           TREONTA ant. The TRENT.
France. DRUENTIA ant., now the DURANCE.
Germany. The DREWENZ. Prussia.
Italy. TRUENTIUS ant., now the TRENTO.
Russia. TURUNTUS ant., now the DUNA.

In the sense of tortuousness I am inclined
to bring in the following, referring them to
Old Norse *meis*, curvatura, Eng. *maze*, &c.
This seems most suitable to the character of
the rivers, as the Maese or Meuse, and the
Moselle. The word seems wanting in the
Celtic, unless we think of the Welsh *mydu*,
to arch, to vault. The other word which
might put in a claim is *mos*, which, in the
sense of marsh, is to be traced both in the
Celtic and German speech, and whence, as
supposed, the name of the ancient Mysia or
Mœsia.

1. *England.* The MAESE. Derbyshire.
   Scotland. The MASIE. Aberdeen.
   France, &c. MOSA, 1st cent. B.C.   The MAAS,
   MAES, or MEUSE.
   Germany. MISS(AHA), 8th cent. The MEISS(AU).
   The MIES in Bohemia.
2.   *With the ending en.*
   Italy. The MUSONE. Two rivers.
3.   *With the ending el.*
   Germany. MOSELLA, 1st cent. The MOSELLE.

The only names which appear to contain
an opposite sense to the foregoing are the

BEINA of Norway, and the BANE of Lincoln-
shire, which seem to be from Old Norse
beinn, North Eng. *bain,* straight, direct.

CHAPTER VII.

QUALITY OF WATERS.

There are a number of river-names in which the sense of clearness, brightness, or transparency is to be traced. From the Sansc. *cand*, to shine, Lat. *candeo*, Welsh, Ir. Arm., and Obs. Gael. *can*, white, clear, pure, we get the following. But the Gael. and Ir., *caoin*, soft, gentle, is a word liable to inter-mix.

1. *England.*	The CANN. Essex.
	The KEN or KENT. Westmorland.
	The KENNE. Devonshire.
Scotland.	The KEN. Joins the Dee.
	The CONN. CONA of Ossian.
	CANDY burn. Lanarkshire.
Wales.	The CAIN. Merioneth.
Germany.	CONE, 9th cent., now the COND.
Russia.	The KANA. Gov. Yeniseisk.
India.	The CANE or KEN—here ?

2.		*With the ending en.*
	Scotland.	The Conan. Dingwall.
	Italy.	The Cantiano. Pont. States.
3.		*With the ending er.*
	England.	The Conder. Lancashire.
		The Conner. Cornwall.
	Switzerland.	The Kander.
4.		*Compounded with vi, wy, river.*
	Wales.	Conovius ant. The Conway.

The Old Celtic word *vind*, found in many ancient names of persons and places, as Vindo, Vindus, Vindanus,* Vindobona, Vindobala, &c., represents the present Welsh *gwyn* (= *gwynd*), and the Ir. *finn* (= *find*), white. "The Celt. *vind*," observes Gluck, "comes from the same root as the Goth. *hveit*; it stands for *cvind* with an intrusive *n*; the root is *cvid* = the Germ. root *hvit*." The meaning in river-names is bright, clear, pure.

1.	*England.*	The Vent. Cumberland. '
		The Quenny. Shropshire.
	Wales.	The Gwynedd (= Gwynd ?)

---

* The three first are names of persons, and to them we might perhaps refer the present family names Window, Windus, Vindin ; though Windo and Winidin were also ancient German names.—(*Förstemann's Altdeutsches Namenbuch.*) The Welsh name Gwyn and the Irish Finn represent the later form of the word.

Ireland.	The FINN.	Ulster.
France.	The VENDÉE.	Dep. Vendée.
Russia.	The ·VIND(AU) or WIND(AU).	

2. *With the ending en.*
Scotland. The FINNAN. Inverness.

3. *With the ending er.*
England. The lake WINDER(MERE) ?*
Ireland. WINDERIUS, *Ptolemy*, a river not identified.

4. *With the ending rn, p.* 34.
Scotland. The FINDHORN. Inverness.

5. *With the ending el.*
England. The WANDLE. Surrey.
Germany. FINOLA, 8th cent., now the VEHNE.

From the Welsh *llwys*, clear, pure, Gael. *las*, to shine, Gael. and Ir. *leus*, light, cognate with Old Norse *lios*, clear, pure, Lat. *luceo*, &c., I derive the following. The Gael. *lâ*, *lo*, day, must, I think, contain the root.

1. *England.* The LIZA. Cumberland.
Scotland. The LOSSIE. Elgin.
France. The LEZ. Dep. Herault.
Belgium. The LESSE.
Germany. The LÖOSE. Pruss. Sax.

2. *With the ending en.*
France. The LIZENA.

---

* Or, as I have elsewhere derived it, from the man's name Winder, still found in the district.

Sweden.	The LJUSNE.	Falls into the Gulf of Bothnia.

3.   *With the ending er.*

Germany.	LESURA, 11th cent., now the LIESER.
	LYSERA, 10th cent., now the LEISER.

From the root of the above, by the prefix *g*, is formed Gael. and Welsh *glas*, blue or green, (perhaps originally rather transparent), and the Old Norse *gladr*, Old High Germ. *glatt*, shining.

Scotland.	The GLASS.	Inverness.
	GLASS.	A lake, Rosshire.
Germany.	The GLATT.	Hohenzollern Sig.
Switzerland.	GLATA, 8th cent.	The GLATT.

Also from the same root come Gael., Ir., and Arm. *glan*, Welsh *glain*, pure, clear, Eng. *clean.*

England.	The GLEN.	Northumberland.
	The GLEN.	Lincolnshire.
	The CLUN.	Shropshire.
France.	The GLANE.	
Germany.	GLANA, 8th cent. The GLAN, two rivers, and the GLON, three rivers.	
Switzerland.	The KLÖN, a small but beautiful lake in the Klönthal—here, or to *klein*, little?	
Italy.	CLANIS ant., now the CHIANA.	
	CLANIUS ant., in Campania.	
Illyria.	The GLAN, in Carinthia.	

From the Old High Germ. *hlutar*, Mod.
Germ. *lauter*, pure, Förstemann derives the
following rivers of Germany.   Hence also
the name of Lauterbrunnen (*brunnen*, fountain), in Switzerland.

> *Germany.*   HLUTR(AHA), 7th cent.   The LAUTER,
> the LUDER, the LUTTER.
> The SOMMERLAUTER in Wirtemburg
> seems to merit the title of pureness
> only in summer.

The following names I think can hardly
be referred to the same origin as the above,
though according to Lhuyd, who derives
them from Welsh *gloew*, clear, and *dwr*,
water, they would have the same meaning.

> *England.*   The LOWTHER.   Westmoreland.
> *Scotland.*   The LAUDER.   Berwickshire.
> *France.*   The LAUTER.

In the Gael. and Ir. *ban*, white, we may
probably find the meaning of the following.

> *Ireland.*   The BANN.   Three rivers.
> *Scotland.*   The BANN(OCK) by Bannockburn.
> *Bohemia.*   The BAN(ITZ).

Of the two following names the former
may be referred to the Welsh *claer*, and the

latter to the Swed. *klar*, both same as Eng. *clear*.

| Ireland. | The CLARE. | Connaught. |
| Sweden. | The KLARA (*å*, river). | |

From the Welsh *têr*, pure, clear, we may get the following. The root is found in Sansc. *tar*, to penetrate, whence *taras*, transparent.

	Italy.	The TARO.	Joins the Po.
	Siberia.	The TARA.	Joins the Tobol.
2.		With the ending en.	
	England.	The TEARNE.	Shropshire.
		The DEARNE.	Yorkshire.
	France.	The TARN.	Joins the Garonne.
3.		With the ending es.	
	Hungary.	The TARISA.	

The following two rivers of Germany may, as suggested by Förstemann, be referred to Old High Germ. *flât*, pure, bright.

1. Germany.	FLAD(AHA), 8th cent. Not identified.
2.	With the ending enz.
Germany.	FLADINZ, 11th cent., now the FLAD-NITZ.

The root *bil* I have, in river-names generally, referred at p. 84 to the Celtic *biol*, water. But in the Slavonic districts we may also think of the Slav. *biala*, white, though we cannot say but that even there the Celtic word may intermix.

Germany.   The BILA in Bohemia.
           The BIALA in Silesia.
Russia.   The BIELAYA.   Joins the Kama.
           The BIALY.   Joins the Narew.

From the Old High Germ. *swarz*, Mod. Germ. *schwarz*, black, are the names of several rivers of Germany, as the SCHWARZA, the SCHWARZAU, the SCHWARZBACH, &c. Also in Norway we have two rivers called SVART ELV, and in Sweden the SVART AN, which falls into the Mälar Lake. From the Old Norse *doeckr*, dark, may be the DOKKA in Norway, but for the DOCKER of Lancashire the Gael. *doich*, swift, may be more suitable.

The Welsh *du*, Gael. *dubh*, black, probably occurs in river-names, but I have taken, p. 36, the meaning of water, as found in Obs. Gael. *dob*, to be the general one. The Welsh *dulas*, dark or blackish blue, is found in the DOWLES of Shropshire, and in several streams of Wales. The DOUGLAS of Lanarkshire shews the original form of the word, from *du*, black, and *glas*, blue.

The root *sal* I have taken at p. 76 to have in some cases the simple meaning of water. But in the following the quality of saltness comes before us as a known characteristic.

Germany.   SALZ(AHA), 8th cent.   The SALZA by Salzburg.

                 SALISUS, 8th cent., now the SELSE.

                 The SALZE. Joins the Werre.

Hungary.   The SZALA.* Falls into Lake Balaton.

Of an opposite character are the following, which we may refer to Welsh *melus*, Gael. and Ir. *milis*, sweet, *millse*, sweetness. Some other rivers, as the ancient MELAS in Asia Minor, now the Kara-su (Black river), and three rivers of the same name in Greece, must be referred to Gr. μελας, black.

Germany.   MILZISSA, 8th cent., now the Mülmisch.

                 MILSIBACH, 11th cent.

Portugal.   MELSUS ant. (Strabo).

---

* The waters of Lake Balaton are described as "slightly salt," and I assume from the name that the Szala is the river from which its saltness is derived.

CHAPTER VIII.

THE SOUND OF THE WATERS.

The GRETA in the English Lake District has been generally derived from Old Norse *grâta*, Scotch *greet*, to weep or mourn, in allusion to the wailing sound made by its waters. There is also a GRETA in Westmoreland and a GRETA BECK in Yorkshire. In the Obs. Gael. and Ir., *greath* also signifies a noise or cry, so that it is quite possible that the original Celtic name may have been retained in the same sense.

Of an opposite meaning to the above is the name BLYTHE of several small rivers in England. I do not see how this can be otherwise derived than from the Ang.-Sax. *blithe*, merry. And how appropriate this is to many of our English streams we hardly need poetic illustration to tell us.

Of a corresponding meaning with the Saxon name Blythe may be the AVOCA or OVOCA of Wicklow, the OBOKA of Ptolemy. Baxter refers it to Welsh *awchus,* acer, a word of no very cheerful association for the spot where

> " Nature has spread o'er the scene
> Her purest of crystal, and brightest of green."

The Gael. *abhach,* blythe, sportive, would seem to give a better etymon for the bright waters of Avoca. Whether the OCKER of Germany (ant. OBOCRA, OVOCRA, OVOKARE), may be derived from the same word I do not know sufficient to judge.

From the Gr. βρέμω, Lat. *fremo,* Ang.-Sax. *bremman,* to roar, Old Norse *brim,* roaring or foaming of the sea, Welsh *ffrom,* fuming, Gael. *faram,* din, I take the following. The following description given by Strabo* of the Pyramus shews the appropriateness of the derivation. "There is also an extraordinary fissure in the mountain,

---

* Bohn's Translation.

U

(Taurus), through which the stream is carried. . . . On account of the winding of
its course, the great contraction of the stream,
and the depth of the ravine, *a noise, like that
of thunder, strikes at a distance on the ears
of those who approach it.*"

1. *England.*	The FROME. Five rivers.
	The FRAME. Dorsetshire.
Germany.	BRAM(AHA) or BREM(AHA), 9th cent., a stream in Odenwald.
	PRIMMA, 9th cent. Near Worms.
	The PRÜM in Prussia.
Denmark.	The BRAM(AUE) in Holstein.
Italy.	FORMIO ant. in Venetia.
Asia Minor.	PYRAMUS ant., now the Jihun.
2.	*With the ending t.*
Germany.	The PFREIMT in Bavaria.
3.	*With the ending nt.*
Germany.	PREMANTIA, 9th cent., now the PRIMS.
4.	*With the ending es.*
Greece.	PERMESSUS ant. Bœotia.

In the Gael. *fuair*, sound, *faoi*, a noisy
stream, we may perhaps find the origin of
the FOWEY in Cornwall, and of the FOYERS
in Inverness, the latter of which is noted as
forming one of the finest falls in Britain.

From the Gael. *gaoir*, din, we may derive the GAUIR in Perthshire ; and from *toirm* of the same meaning, perhaps the TERMON in Ulster.   Hence might also be the TROME and the TRUIM, elsewhere derived at p. 70.

From the Gael. *durd, durdan,* Welsh *dwrdd,* humming or murmur, Lhuyd derives the name DOURDWY, of some brawling streams in Wales ; but quoting the derivations of some other writers, he adds, with more humility than authors generally possess—" Eligat Lector quod maxime placet." To the same origin may probably also be referred the DOURDON in France, Dep. Seine-Inf.

CHAPTER IX.

JUNCTION OR SEPARATION OF STREAMS.

There are several river-names which contain the idea, either of the junction of two
streams, or of the separation of a river into
two branches.   The Vistula, Visula, or
Wysla, (for in these various forms it appears
in ancient records), is referred by Müller,*
rightly as I think, to Old Norse *quisl*, Germ.
zwiesel, branch, as of a river.   A simpler
form of *quisl* is contained in Old Norse *quistr*,
ramus, and the root is to be found in Sansc.
dwis, to separate, Gael. and Ir. *dis*, two.   The
Old Norse name of the Tanais or Don, according to Grimm (*Deutsch. Gramm.* 3, 385),
was Vana-quisl.   The word *whistle*, found as
the ending of some of our local names, as
Haltwhistle in Northumberland, and Osbald-

---

* Die marken des Vaterlandes.

whistle in Lancashire, I take to be = the Old
Norse *quisl :* the sense might be that of the
branching off of two roads or two streams.
In an account of the hydrography of Lanarkshire, for which I am indebted to the kind-
ness of a Friend, there is a burn called Gala-
whistle, which compares with the above Old
Norse Vana-quisl. In connection with the
Vistula Jornandes introduces a river Viscla,
which has been generally considered to be
merely another form of the same word—
Reichard* being, as I believe, the only writer
who considers it to be a different river. It
seems to me a curious thing that it has never
occurred to any one to identify it with the
Wisloka, which joins the Vistula near Bara-
nov. The modern name must contain the
correct form, for Wisloka = an Old High
Germ. Wisilacha, from *acha* or *aha*, river,
and is the same as the Wisilaffa or Wislauf,
from *afa* or *apa*, river. The following names
I take to be all variations of the same word.

---

* Germanien unter den Römern.

1. *France.*   The OUST.   Dep. Côtes-du-Nord.
   Germany.   The TWISTE.   Joins the Diemel.
   The QUEISS.   Pruss. Silesia.
   Russia.   The UIST.   Joins the Tobol.
   The USTE.   Joins the Dwina.

2.   *With the ending en.*
   Germany.   QUISTINA, 11th cent., now the KÖSTEN.

3.   *With the ending er.*
   France.   The VISTRE.   Dep. Gard.
   Belgium.   The VESDRE.   Joins the Ourt.
   Germany.   The VEISTR(ITZ).   Pruss. Silesia.

4.   *With the ending rn.*
   Germany.   QUISTIRNA, 8th cent., now the TWISTE,
   joins the Oste.

5.   *With the ending el=O. N. quisl.*
   Germany,   VISTULA, 1st cent., Germ. WEICHSEL.
   &c.   WISL(OKA), joins the Vistula. (*See
                above.*)
   The WISL(OK).   Joins the San.
   WISIL(AFFA), 11th cent., now the
   WISL(AUF).
   France.   The VESLE.   Joins the Aisne.

The following seem also to contain the
Germ. *zwei,* Eng. *two,* and to have something
of a similar meaning to the foregoing.

1. *Germany.*   The ZWITT(AWA) or ZWITT(AU) in
   Moravia.

2.   *With the ending el.*
   Germany.   The ZWETTEL in Austria.

I include also here the SCHELDT or SCHELDE, (the SCALDIS of Cæsar), which I think is to be explained by the Old Norse *skildr*, Dan. *skilt*, separated, in allusion to the two mouths by which it enters the North Sea. And to the same origin may be also placed the SCHILT(ACH) of Baden, which falls into the Kinzig.

From the Gael. *caraid*, duplex, may probably be the two CARTS in the County of Renfrew, the united stream of which enters the Firth of Clyde near Glasgow.

CHAPTER X.

BOUNDARY OR PROTECTION.

The idea of a river as a protection or as a boundary seems to indicate a more settled state of society, and therefore not to belong to the earliest order of nomenclature. And consequently, though this chapter is not quite so bad as the well-known one "Concerning Owls," in Horrebow's Natural History of Iceland, the sum and substance of which is that "There are no owls of any kind in the whole Island"—it will be seen that the number of names is very small in which such a meaning is to be traced.

The word *gard*, which in the Celtic, Teutonic, Slavonic, and other tongues has the meaning of protection or defence, must, I think, have something of the same meaning in river-names. Or it may perhaps rather

be that of boundary, for the two senses run
very much into each other.

1. *France.*   The GARD.   Joins the Rhone.
   Germany.   GARD(AHA), 8th cent.   The GART-
            (ACH).
            The KART(HAUE) in Prussia.
2.   *With the ending en.*
   Scotland.   The GAIRDEN.   Joins the Dee.
   France.   The GARDON.   Joins the Rhone.
   Greece.   JARDANUS ant. in Crete—here?

In the Gael. *sgia*, Welsh *ysgw*, guard, protection, and in the Welsh *ysgi*, separation or
division, we have two senses, of which the
latter may be more suitable for the following. The Editor of Smith's Ancient Geography suggests that the Scius of Herodotus
may be the present Isker in Bulgaria : in an
etymological point of view this seems probable, for as Scius = Welsh *ysgi*, so Isker =
Welsh *ysgar* of the same meaning.

Netherlands.   The SCHIE by Schiedam.
Danub. Prov.   SCIUS ant., now the ISKER?

From the Gael. *scar, sgar*, Welsh *ysgar*,
Ang.-Sax. *scêran*, to divide, in the sense of
boundary, may be the following. The small

river Scarr in Dumfriesshire forms for six
miles a boundary between different parishes.*

  1. *England.*   The SHERE.   Kent.
      Scotland.   The SCARR.   Dumfriesshire.
               The SHIRA.   Argyle.
      Germany.   SCERE, 11th cent.   The SCHEER.
  2.   *With the ending en.*
      England.   The SKERNE.   Durham.
      Germany.   SCHYRNE, 11th cent., not identified.

Any names in which the sense of *land*,
terra, occurs, may, I think, be explained most
reasonably in the sense of boundary or ter-
ritorial division. To this Grimm places the
FULDA of Germany, FULD(AHA), 8th cent.,
referring it to Old High Germ. *fulta*, Ang.-
Sax. *folde*, earth, ground.

Perhaps also to a similar origin may be
referred the MOLD(AU) in Bohemia, and the
MOLD(AVA) of Moldavia. But the Gael.
and Ir. *malda*, *malta*, gentle, slow, Anglo-
Sax. *milde*, Eng. *mild*, may be perhaps more
suitable : the MULDE, which joins the Elbe,
and which in the 8th cent. appears as MILDA,
seems more probably from this origin.

---

* Statistical account of Scotland.

The BORD(AU), formerly BORDINE, which forms for some distance the boundary between East and West Friesland, may, as suggested by Förstemann, be derived from Old Fries. and Anglo-Saxon *bord*, border. Another river of the same name (p. 33) may perhaps be otherwise derived.

I am inclined to bring in here the GRANTA, and to suggest that it may have been a Sax. or Angle name of the Cam, or of a certain part of the Cam. This river seems to have formed one of the boundaries of the country of the Gyrvii;* its name appears in Henry of Huntingdon as Grenta; and the Old Norse *grend*, Mod. Germ. *grenze*, boundary, seems a probable etymon.

---

* See an article by the Rev. W. Stubbs on "The Foundation and early Fasti of Peterborough," in the Archæological Journal for Sept., 1861.

CHAPTER XI.

In this chapter I include some names which do not come under any of the foregoing heads, or which have been omitted in their places.

The following have generally been referred to Gael. *caol*, straight, narrow.

1. *England.*　The Cole.　Warwickshire.
　　　　　　　　The Coly.　Devon.
2. 　　　　　　　*With the ending en.*
　England.　The Colne.　Three rivers.

But even if this derivation is to be received, we must seek another meaning for the Kola in Russian Lapland, and the Koli(ma) in Siberia—the latter in particular being a large river, with a wide estuary.

The Gael. and Ir. *beag*, little, forms the ending of some Irish river-names, as the Awbeg, the Owenbeg, and the Arobeg.* The

---

* The derivation at p. 120 I must retract, finding *beg* as a termination of other Irish river-names.

meaning in all these cases is "little river"—
owen being the same as *avon, aw* the simple
form *av* of the same word, and *aro* an appellative as at p. 38, now lost in the Celtic.

From the Gael. *suail*, small, have also
been derived the Swale and other following
rivers. Chalmers rightly objects to this as
inconsistent with the character of the rivers,
though the derivation which he proposes to
substitute, from *ys-wall*, a sheltered place,
affords, it must be admitted, no very happy
alternative. I think the word contained
must be related to Old High German *swal*,
Old Norse *svelgr*, gurges, Eng. *swell*, though
it is wanting in the Celtic.

1. *England.* The SWALE. Two rivers, Kent and
Yorkshire.
The SWILY. Gloucestershire.
Ireland. The SWELLY. Donegal.
The SWILLY. Ulster.
Germany. SUALA ant. The SCHWALE.
France. SULGAS ant., now the Sorgue.
Russia. The SULA—here ?

2. *With the ending en.*
Ireland. The SULLANE.

The following must be referred to Old High Germ. *sualm*, gurges, an extension of the previous word *sual*.

Germany.	SUALMAN(AHA), 8th century. The SCHWALM.
	SULMANA, 8th cent. The SULM.
Belgium.	The SALM. Prov. Liège.
France.	The SOLMAN. Dep. Jura.

The Shannon has by some writers been derived from Ir. *sean* or *shean*, old. But inasmuch as there is no river that is otherwise than old, the term could only be used in a poetic sense, like "that ancient river, the river Kishon." A more suitable etymon, however, seems to me to be found in Ir. and Obs. Gael. *siona*, delay; this corresponds with the Gaelic form of the name, Sionan, given by Armstrong.

Scotland.	The SHIN. Sutherland.
Ireland.	SENUS (Ptolemy). The SHANNON.
Germany.	SINNA, 8th cent. The SINN.
Belgium.	The SENNE. Joins the Dyle.
Italy.	SENA ant., now the Nevola.
Aust. Pol.	The SAN, two rivers—here ?
India.	The SEENA—here ?

From the Gael. *cobhair*, Ir. *cubhair*, foam, froth, appear to be the following.

England.	The COBER.	Cornwall.
	The COVER.	Yorkshire.
Russia.	The CHOPER.	
Asia.	CHABORAS ant., now the KHABUR—here ?	
India.	CHABERIS ant., now the CAVERI—here ?	

From the Ir. and Obs. Gael. *breath*, pure, clear, I take to be the following.

England.	The BRATHA.	Lake District.
Scotland.	The BROTH(OCK).	Forfar.
Germany.	The BRETT(ACH).	Joins the Kocher.
	The BRAT(AWA) in Bohemia.	
	BRAHT(AHA),* 10th century. The BRACHT—here ?	
Asia Minor.	PRACTIUS ant.—here ?	

And from the Ir. *brag*, running water, I follow Mone in taking the following.

1.	*England.*	The BRAY.	Devon.
	Ireland.	The BRAY.	Wicklow.
	France.	The BRAY.	Joins the Loire.
	Germany.	The BREGE, in the Scharwarzwald.	
2.		*With the ending en.*	
	England.	The BRAINE.	Joins the Blackwater.
	Ireland.	BREAGNA, an old name for the Boyne.	

---

* Wiegand, (Oberhessische ortsnamen), refers this name to Old High Germ. *braht*, fremitus.

A root for river-names, to which might be put the following, is found by Förstemann in Old High Germ. *rôr*, Mod. Germ. *rohr*, arundo, Eng. *rush*.

> *Germany.* ROR(AHA), 11th century, now the ROHRBACH.
> RURA, 8th cent. The RUHR.
> *Holland.* The ROER. Joins the Maas.

The word *sil* in river-names would seem to have the meaning of still or sluggish water. The Gael. has *sil*, to drop, rain, drip; and the Arm. has *sila*, to filter. (The Old Fries. *sil*, canal, seems hardly a related word; it appears more probably to be connected with Old Norse *sila*, to cut, to furrow.) According to Pliny, the Scythian name of the Tanais or Don was Silis; and several other Scythian rivers had the same name, (*Grimm, Gesch. d. Deutsch. Sprach.*) In this point of view the above derivation might seem too restricted, and we might think of *sil*, as of *sal*, (p. 75), as formed by the prefix *s* from the root *al* or *il*, to go, (p. 71,) in the simple meaning of water. According to Strabo and

Pliny the Silaris òf Italy had the property of petrifying any plant thrown into it ; but as, according to Cluvier, the modern inhabitants of its banks know nothing of any such property, it would rather seem as if the story had been made to fit the supposed connection of the name with *silex*, flint.

1. *Switzerland.* SIL(AHA), 11th cent. The SIHL.
   Italy. SILIS ant., now the SILE.
   Scotland. The SHIEL in Argyleshire—here ?
   Germany. The SCHYL (ant. Tiarantus)—here ?
2. *With the ending en.*
   Sweden. SILJAN. Lake.
   Russia. The SHELON—here ?
3. *With the ending er.*
   Naples. SILARIS ant., now the SILARO.

The form *silv* I take to be an extension of *sil*, similar to others previously noticed.

1. *Russia.* The SILVA. Gov. Perm.
2. *With the ending er.*
   England. The SILVER. Devon.

The SIMOIS in the Plain of Troy I have suggestively placed at p. 119 to Gael. *saimh*, slow, tranquil. But, taking the epithet *lubricus* applied to it by Horace, we might

w

perhaps seek a stronger sense from the same root, as found in Welsh *seimio*, to grease, *saim*, tallow.

The water of the LIPARIS in Cilicia, according to Polyclitus, as quoted by Pliny, was of such an unctuous quality that it was used in place of oil. Probably only for the purpose of anointing the person, to which extent the story is confirmed by Vitruvius. Hence no doubt its name, from Sansc. *lip*, to be greasy, Gr. λιπαρος, unctuous.

Grimm *(Gesch, d. Deutsch. Sprach.)* suggests a similar origin for the Ister, p. 117, referring it to Old Norse *istra*, Dan. *ister*, fat, grease, Gr. στέαρ. He puts it, however, in a metaphorical sense, as " the fattening, fructifying river." With deference, however, to so high an authority, this explanation seems to me rather doubtful. For the ending *ster*, as I have elsewhere observed, is common to many river-names, and I have taken it to be, like the Arm. *ster*, formed by a phonetic *t*, from the Sansc. *sri*, to flow.

Also, from the root of the Sansc. *sri*, to flow, I take to be Gael. *sruam*, and again taking the phonetic *t*, the word *stream*, *strom*, common to all the Teutonic dialects. In these two forms we find the ancient names of two rivers—the SYRMUS of Thrace, and the STRYMON or STRUMON, the present STRUMA, of Macedonia.

CHAPTER XII.

CONCLUSION.

The names of rivers form a striking commentary on the history of language, so admirably expounded to the general reader in the recent work of Professor Max Müller.

When we review the long list of words that must have once had the meaning of water or river, we can hardly fail to be struck with the number that have succumbed in what he so aptly terms "the struggle for life which is carried on among synonymous words as much as among plants and animals."

We see too how large a portion of this long list of appellatives may ultimately be traced back to a few primary roots. And how even these few primary roots may perhaps be resolved into a still smaller number of yet more simple forms.

I take for instance, as a primitive starting
point in river-names, the Sansc. root *i, á*, or
ay, signifying to move, to flow, to go. We
have appellatives even in this simple form,
as the Old Norse *á*, Anglo-Sax. *ǽ*, water,
river. But whether they directly represent
the root, or whether, like the French *eau*, p.
30, they have only withered down to it again,
after a process of germinating and sprouting,
I do not take upon me to determine.

Then we have the roots, also of the kind
called primary, *ab, ar, ir, ag, ikh, il, it*, all
having the same general meaning, to move,
to go, and from which, as elsewhere noticed,
are also derived a number of appellatives for
water or river in the various Indo-European
languages. I should be inclined to suggest
that the whole of these are formed upon, and
are modifications of the simple root *i, á*, or
ay, and that the following remarks made by
Max Müller respecting secondary roots, may
be extended also to them. "We can frequently observe that one of the consonants,

in the Aryan languages, generally the final,
is liable to modification.   The root retains
its general meaning, which is slightly modified and determined by the changes of the
final consonants." He instances the Sansc.
tud, tup, tubh, tuj, tur, tuh, tus, all having the
same general meaning, to strike.

Again—there are forms such as *ang, amb,
and,* &c., which are merely a strengthening
of the roots *ag, ab, ad,* or *at,* and which also
are found in a number of appellative forms.

We might pursue the subject still further,
and enquire whether the secondary forms,
such as *sar, sal, car, cal,* all having the same
general meaning, to move, to go, may not be
formed, by the prefix of a consonant, on the
roots *ar* and *al,* and so also be ultimately
referred to the simple root *i* or *d.*

As also the silent and ceaseless flow of
water is the most natural and the most common emblem of the efflux of time ; so in the
same root is to be found the origin of many
of the words which mean time and eternity.

The Gr. *αει*, the Goth. *aiv*, the Anglo-Sax. *awa*, Eng. *ever* and *aye*, are all from this same root, so widely spread in river-names, and express the same idea which speaks—

> " For men may come, and men may go,
> But I go on for ever."

P. 25.

To the root *ab* or *ap*, water, place the Lith. and Lett. *uppe*, river, whence the following.

Germany. The OPPA in Silesia.
Russia. The UPA. Joins the Oka.
The UFA. Joins the Bielaya.

P. 33.

To the root *ud* place as an appellative the Obs. Gael. *ad*, water. And add to form No. 1 the following names.

Russia. The UDA. Gov. Kharkov.
France. The ODDE. Dep. Allier.

P. 35.

The Celt. word *and* or *ant*, water, is nothing more than a strengthening of the above Obs. Gael. *ad*.

P. 40.

In referring to the root *ark*, *erk*, I have omitted the Ir. *earc*, water, the appellative most nearly concerned. The Basque *erreca*, brook, might be taken to be borrowed from the Celtic, did we not find in the same language the more primitive words *ur* and *errio*, p. 38, which seem to form a link with the Indo-European languages.

P. 49.

To the root *nig, ni*, place—
1. *France.*   The Né.   Joins the Charente.
   Norway.   The Nia.   Stift Trondjem.
3.   *With the ending es.*
   Russia.   The Nerussa.   Gov. Orel.

P. 63.

To the root *wig, wic, wy*, place the two following names.   The Welsh *gwy*, water, is the word most nearly concerned in most of the group.

England.   The Wyck.   Buckinghamshire.
Russia.   The Ui.   Gov. Orenburg.

P. 64.

To the root *vip* place as an appellative the Welsh *gwibio*, to rove, wander, *gwibiau*, serpentine course. Probably upon the whole the sense of tortuousness is that which should be recognized.   The following name probably belongs to form No. 1.

Spain.   The Quipar.   Joins the Segura.

P. 70.

The Celtic languages have a trace of the word *trag*, to run, in the Old Ir. *traig*, foot (*Zeuss, Gramm. Celt.*)

x

P. 83.

For
>*Greece.* PYDARAS ant. Thrace.

Read
>*Thrace.* PYDARAS ant.

P. 84.

To the Ir. *biol*, *buol*, water, place the following names.

>*England.* The BEAULIEU, also called the Exe, in Hampshire.
>*Scotland.* The BEAULY. Inverness.
>*Italy.* PAULO ant., now the Paglione.

P. 85.

I apprehend that in the opinion of Celtic scholars of the present day the Ancient British deity Cocidis is not considered to have any connection with the river Coquet.

P. 91.

It seems probable that the word *asp* in river-names is formed by metathesis from the word *aps*, p. 27, form 5.

P. 97.

The GRYFFE and the GIRVAN may perhaps be better derived from the Gael. *grib*, swift.

P. 132.

To the root *pad* or *pand*, to spread, may probably be placed—

>*England.* The PANT. Essex.

P. 135.

From the root *tan* may be derived the DNIESTER, (= Danaster), from *ster*, river. Or it might be from the root *dan*, as in Danube, p. 116.

P. 136.

The Dan. *tang*, sea-weed, does not seem to be connected with any word signifying water : it represents the Old Norse *tag*, twigg.

P. 145.

To the root *vind*, white, clear, place —

England. The WENTE. Yorkshire.

P. 149.

To the Sansc. *taras*, Welsh *têr*, pure, clear, place—

Thrace. TEARUS ant.

INDEX.

(Ancient Names in Italics.)

Aa, 28
Aach, 28
Aar, 39
Abana, 26
Acaris, 81
Achaze, 31
Adda, 34
Adenau, 34
Adour, 34
Adur, 34
Aenus, 27
Agger, 81
Aghor, 81
Agri, 81
Ahr, 39
Ahse, 31
Ain, 135
Aisne, 31
Aiss, 81
Aiterach, 35
Alass, 75
Alaunus, 71
Alb, 73
Albegna, 74
Alben, 74
Albla, 74
Albula, 74
Alces, 104
Aldan, 72
Alde, 72
Alf, 73
Alhama, 130
Alise, 75
Alisna, 75

Allan, 71
Alle, 71
Aller, 71
Allia, 71
Allier, 74
Allow, 71
Alm, 130
Alma, 130
Alme, 130
Almelo, 130
Almo, 130
Alne, 71
Alpheus, 74
Alpis, 73
Alt, 72
Alta, 72
Alten, 72
Altmühl, 104
Alum Bay, 130
Alz, 75
Amasse, 29
Ambastus, 29
Amber, 29
Amble, 29
Amblève, 29
Amele, 29
Ammer, 29
Amnias, 26
Amon, 26
Andelau, 36
Andelle, 36
Angel, 81
Angera, 81
Angerap, 81

Angrus, 81
Anitabha, 35—Note.
Anker, 81
Annas, 27
Ant, 35
Anton, 36
Anza, 27
Appelbach, 26
Apsarus, 27—Note.
Apsus, 27
Arabis, 120
Aragon, 41, 176
Arak, 41, 176
Arar, 117
Aras, 78
Araxes, 78
Arc, 41, 176
Arga, 41, 176
Argen, 41, 176
Arius, 56
Ariminus, 122
Arke, 41, 176
Arl, 40
Arly, 40
Arme, 122
Armine, 122
Arno, 40
Arobeg, 164
Arosis, 78
Arques, 41
Arrabo, 120
Arrow, 39
Arsia, 78
Arun, 39

Arva, 109
Arve, 109
Ascania, 31
Ash, 31
Asopus, 92, 178
Aspe, 92, 178
Astura, 58
Au, 28
Aube, 73
Aulne, 71
Aune, 27
Aupe, 73
Aurach, 39
Auray, 39
Auve, 74
Aven, 26
Avia, 25
Aviz, 27
Avoca, 153
Avon, 26
Avre, 26
Awbeg, 164
Awe, 28
Axe, 30
Axius, 31
Axona, 31
Axus, 31

Bahr, 65
Bandon, 132
Bane, 143
Banitz, 148
Bann, 148
Bannock, 148
Bar, 65
Barrow, 65
Baunach, 84
Beaulieu, 178
Beauly, 178
Beela, 84
Behr, 65
Behrun, 65
Beina, 143
Beraun, 65
Bere, 65
Berre, 65

Beuvron, 84
Bever, 84
Biala, 150
Bialy, 150
Biberbach, 84
Bibra, 84
Bielaya, 150
Bièvre, 83
Bila, 150
Billæus, 85
Binoa, 82
Birse, 101
Blythe, 152
Bode, 132
Boderia, 132
Bogen, 138
Bogie, 138
Bolbec, 85
Bollaha, 85
Bord, 133
Bordau, 163
Bowe, 138
Boyle, 85
Boyne, 84
Bracht, 167
Braine, 167
Bramaue, 154
Bratawa, 167
Bratha, 167
Bray, 167
Breagna, 167
Brege, 167
Bresle, 101
Brettach, 167
Brosna, 101
Brothock, 167
Bucket, 138
Bug, 138
Buhler, 85
Buller, 85
Bullot, 85
Burzen, 101

Cædrius, 108
Cailas, 110
Cain, 144

Calbis, 113
Caldhowa, 112
Calder, 112
Caldew, 112
Callan, 110
Callas, 110
Callipus, 113
Calore, 110
Calpas, 113
Cam, 138
Candy Burn, 144
Cane, 144
Cann, 144
Cantiano, 145
Caresus, 114
Carpino, 97
Carpis, 97
Carron, 139
Cart, 159
Caveri, 167
Cayster, 68
Celadon, 112
Celydnus, 112
Cerbalus, 98
Cersus, 114
Cestrus, 68
Chalus, 110
Chalusus, 110
Cham, 138
Char, 139
Charente, 139
Chares, 139
Chelt, 112
Chelva, 113
Cher, 139
Chiana, 147
Chiers, 114
Choaspes, 68, 178
Choper, 167
Chor, 139
Churne, 139
Cladeus, 80
Clanius, 147
Clare, 149
Cleddeu, 79
Clitora, 80

Clitumnus, 80
Cloyd, 79
Cludros, 80
Clun, 147
Clwyd, 79
Clyde, 79
Cober, 167
Cocbróc, 86
Cocker, 86
Cockley-beck, 87
Cocytus, 87
Coker, 86
Colapis, 113
Cole, 164
Colne, 164
Coly, 164
Conan, 145
Cond, 144
Conder, 145
Conn, 144
Conner, 145
Conway, 145
Coquet, 87
Coralis, 139
Cover, 167
Cremera, 140
Cremisus, 140
Crummock, 140
Cuckmare, 87
Curalius, 139
Cydnus, 108
Cyrus, 139

Dahme, 135
Dalcke, 106
Dane, 135
Danube, 116
Daradax, 105
Daradus, 105
Darme, 70
Daubrawa, 37
Deane, 135
Déaume, 135
Dee, 134
Deel, 105
Delvenau, 106

Demer, 135
Derwent, 141
Desna, 107
Deva, 135
Dill, 105
Dillar Burn, 106
Dista, 107
Dive, 135
Dniester, 179
Dobur, 37
Docker, 150
Dodder, 90
Dokka, 150
Dommel, 90
Don, 135
Donge, 136
Dora, 37
Dordogne, 38
Doubs, 36
Douglas, 150
Dourdon, 155
Dourdwy, 155
Douro, 37
Doux, 36
Dove, 36
Dovy, 36
Dow, 36
Dowles, 150
Drac, 70
Drage, 70
Drammen, 70
Dran, 69
Drave, 69
Drewenz, 141
Drome, 70
Drone, 69
Dronne, 69
Dubissa, 37
Duddon, 90
Dude, 90
Durance, 141
Durme, 70
Durra, 37
Dussel, 107
Duyte, 90
Dyle, 106

Earne, 40
Ebrach, 26
Ebro, 26
Ecolle, 69
Eden, 35
Eder, 34
Edrenos, 34
Eem, 28
Eger, 81
Ehen, 27
Eichell, 28
Eider, 35
Eisach, 32
Eitrach, 35
Elbe, 73
Eld, 72
Elda, 72
Ellé, 71
Ellen, 71
Ellero, 71
Ellison, 75
Elvan, 74
Elz, 75
Emba, 29
Emele, 29
Emme, 28
Emmen, 29
Emmer, 29
Ems, 29
Ens, 27
Era, 39
Erens, 138
Erft, 40
Ergers, 41
Erl, 40
Erla, 40
Erms, 122
Erpe, 109
Erve, 109
Eschaz, 31
Esk, 31
Eskle, 31
Esla, 33
Esque, 31
Ettrick, 35
Eure, 34

Evan, 26
Evenus, 26
Eye, 28
Eypel, 27
Exe, 31

Fal, 130
Feale, 130
Fillan, 130
Fils, 130
Findborn, 146
Finn, 146
Finnan, 146
Fladaha, 149
Fladnitz, 149
Fleet, 66
Flieden, 66
Flietnitz, 66
Flisk, 67
Foilagh, 130
Formio, 154
Forth, 115
Fowey, 154
Foyers, 154
Frame, 154
Fraw, 115
Frome, 154
Froon, 115
Fulda, 162

Gada, 108
Gaddada, 109
Gade, 108
Gader, 108
Gadmen, 109
Gail, 110
Gairden, 161
Gala, 110
Galthera, 112
Gande, 108
Ganges, 68
Gangitus, 68
Gard, 161
Gardon, 161
Garfwater, 97
Garonne, 13, 114

Garrhuenus, 113
Garry, 113
Gartsch, 161
Garza, 114
Gata, 108
Gauir, 155
Geisa, 108
Gela, 110
Gelt, 112
Geltnach, 112
Geranius, 114
Geron, 114
Gers, 114
Gidea, 108
Giesel, 109
Giesbach, 108
Gingy, 68
Giron, 114
Girvan, 97, 178
Glan, 147
Glass, 147
Glatt, 147
Glen, 147
Glon, 147
Glyde, 80
Gose, 108
Gotha, 108
Gouw, 68
Grabow, 97
Granta, 163
Gravino, 97
Greta, 152
Grumbach, 140
Gryffe, 97, 178
Gwynedd, 145
Gyndes, 108

Haase, 100—Note.
Haliacmon, 104
Halycus, 104
Halys, 75
Hamel, 29
Hamps, 29
Harpa, 109
Harpasus, 109
Hebrus, 26

Helisson, 75
Helme, 130
Helpe, 74
Herk, 41, 176
Hesper, 92, 178
Hespin, 91
Hesudros, 33
Hisscar, 32
Hörsell, 78
Hull, 89
Humber, 29
Hunte, 100
Hypanis, 26
Hypius, 26
Hypsas, 27

Iberus, 26
Idle, 35
Igla, 69
Iglawa, 69
Ihna, 27
Ik, 69
Ilach, 71
Ilavla, 74
Ile, 71
Ilen, 71
Ilek, 104
Ilissus, 75
Ill, 71
Ille, 71
Iller, 71
Illim, 130
Ilm, 130
Ilmen, 130
Ilmenau, 130
Ilse, 75
Ilz, 75
Inda, 23
Inde, 23
Indus, 23
Indre, 23
Ingon, 81
Ingul, 81
Inn, 27
Inney, 27
Ionne, 69

Ipf, 26
Ipoly, 27
Ips, 27
Irat, 138
Irati, 138
Irghis, 41
Irk, 41
Irkut, 41
Irt, 138
Irthing, 138
Irvine, 109
Isac, 31
Isar, 33
Ischl, 31
Ise, 32
Isen, 32
Isère, 32
Isis, 33
Isla, 33
Isker, 161
Ismenus, 33
Isolê, 33
Isper, 92
Isset, 33
Issus, 32
Ister, 33, 117, 170
Itchen, 69
Iton, 35
Itz, 35
Ive, 25
Ivel, 26

Jactus, 100
Jaghatu, 100
Jahde, 100
Jahnbach, 68
Jardanus, 161
Jaxt, 100
Jesmen, 89
Jessava, 89
Jetza, 89
Jezawa, 89
Jisdra, 89
Joss, 89
Jug, 100

Kalitva, 112
Kam, 139
Kama, 139
Kamp, 138
Kana, 144
Kander, 145
Karthaue, 161
Kela, 110
Kelvin, 113
Kemi, 139
Kenne, 144
Kent, 144
Kerr, 139
Kersch, 114
Khabur, 167
Khankova, 68
Klara, 149
Klodnitz, 80
Klön, 147
Kocher, 86
Kohary, 86
Kohlbach, 113
Kokel, 86
Kola, 164
Kolima, 164
Korol 139
Koros, 114
Kösten, 158
Krems, 140
Kroma, 140
Krumbach, 140
Kuchelbach, 87
Kulpa, 113
Kur, 139

Lagan, 45
Lahn, 45
Laimaha, 128
Laine, 45
Laith, 46
Lama, 128
Lambro, 129
Lamme, 128
Lammer, 129
Lamone, 129
Lamov, 128

Lamus, 129
Laucha, 45
Lauder, 148
Lauter, 148
Lave, 45
Lavino, 45
Leach, 44
Leam, 128
Lech, 44
Leck, 44
Lee, 44
Leen, 44
Legre, 44
Leiser, 147
Leith, 46
Leitha, 46
Leithan, 47
Leman, 129
Leman (Lake), 129
Lempe, 128
Lesse, 146
Lethœus, 47
Leven, 45
Lez, 146
Lid, 46
Lida, 46
Lidden, 47
Liddle, 47
Lieser, 147
Liffar, 46
Liffey, 46
Ligne, 44
Lima, 128
Limen, 129
Limmat, 129
Limyrus, 129
Liparis, 170
Lipka, 46
Lippe, 46
Liver, 46
Liza, 146
Lizena, 146
Ljusne, 147
Lloughor, 45
Loing, 45
Loire, 44

Y

Loiret, 14
Lomond (Loch), 129
Looe, 45
Loony, 45
Loose, 146
Lossie, 146
Lot, 72
Loue, 45
Louga, 45
Lougan, 45
Louven, 45
Lowna, 45
Lowther, 148
Luder, 148
Lug, 45
Lugan, 45
Lugano (Lake), 45
Lugar, 45
Lühe, 44
Lune, 45
Lutter, 148
Lye, 44
Lyme, 129
Lyon, 44
Lys, 44

Maas, 142
Macestus, 61
Madder, 88
Madel, 88
Maese, 142
Magra, 60
Mahanuddy, 60
Maia, 60
Malg, 60
Main, 60
Maina, 60
March, 61
Mare, 62
Marecchia, 62
Mark, 61
Marne, 88
Marosch, 62
Marsyas, 62
Masie, 142
Maask (Lake), 62

Matrinus, 88
Matrona, 88
Maw, 60
Mawn, 60
May, 60
Mayenne, 127
Meal, 61
Mede, 88
Medemelacha, 126
Medinka, 126
Medoacus, 127
Medofulli, 126
Medvieditza, 127
Medway, 126
Medwin, 127
Megna, 60
Mehaigne, 60
Mehe, 88
Meissau, 142
Melsus, 151
Meon, 60
Mergui, 62
Mersey, 62
Metauro, 88
Metema, 127
Meuse, 142
Mhye, 60
Midou, 126
Miele, 61
Mies, 142
Milisibach, 151
Moder, 88
Moldau, 162
Moldava, 162
Mora, 61
Morava, 61
Morge, 61
Mörn, 62
Moselle, 142
Moskva, 62
Mourne, 62
Moy, 60
Moyne, 60
Muhr, 61
Mulde, 162
Mülmisch, 151

Muotta, 102
Murg, 61
Murr, 61
Murz, 62
Musone, 142
Muthvey, 102

Naab, 50
Naaf, 50
Nabalis, 51
Nabon, 50
Nahe, 50
Nairn, 49
Namadus, 52
Naparis, 50
Nar, 49
Narenta, 49
Narew, 49
Naron, 49
Narova, 49
Narra, 49
Natisone, 88
Nave, 50
Naver, 50
Navia, 50
Ne, 177
Neagh, (Lake) 49
Neath, 54
Neda, 54
Neers, 49
Neisse, 51
Nenagh, 49
Nene, 49
Nenny, 49
Nent, 49
Nera, 49
Nerja, 49
Nerussa, 177
Ness, 51
Neste, 51
Nestus, 51
Nethan, 54
Nethe, 54
Neutra, 88
Neva, 50
Never, 50

Nevis, 51
Nai, 177
Nia, 49
Nidd, 54
Nidder, 54
Nied, 54
Niemen, 50
Nievre, 50
Nisi, 51
Nissava, 51
Nith, 54
Nive, 50
Nivelle, 50
Noain, 88
Nodder, 88
Noraha 49
Nore, 49
Now, 49
Oarus, 30
Ock, 28
Ocker, 153
Odde, 176
Odder, 34
Oder, 34
Odon, 34
Oenus, 27
Oertze, 78
Ohm, 26
Ohre, 39
Ohrn, 40
Oich, 28
Oikell, 28
Oise, 32
Oka, 28
Oke, 28
Olle, 72
Olmeius, 130
Oltis, 72
Ombrone, 29
Oppa, 176
Orb, 109
Ore, 39
Orge, 41
Orla, 40
Orlyava, 40
Orlyk, 40

Orre, 40
Orrin, 40
Orsinus, 78
Orvanne, 109
Œscus, 31
Oskol, 31
Otter, 34
Ource, 78
Ourcq, 41
Ourt, 138
Ousche, 32
Oust, 158
Owenbeg, 164
Ovoca, 153
Oxus, 31

Paar, 65
Pader, 132
Padus, 132
Palme, 67
Pant, 178
Pantanus, 132
Parde, 133
Parret, 83
Parthenius, 133
Pathissus, 132
Paulo, 178
Pebrach, 84
Pedder, 83
Peen, 81
Peffer, 83
Pelym, 67
Peneus, 82
Penjina, 82
Penk, 82—*Note.*
Penner, 82
Penza, 82
Permessus, 154
Pernau, 65
Persante, 101
Petteril, 83
Pever, 83
Pfreimt, 154
Piana, 82
Piave, 65
Piddle, 82

Pina, 82
Pinau, 82
Pendar, 83
Pindus, 82
Pinega, 82
Pinka, 82
Pitrenick, 83
Plaine, 65
Plau, 65
Plan-see (Lake), 66
Pleiske, 67
Pleisse, 66
Pleistus, 66
Pliusa, 66
Ploen (Lake), 66
Plone, 66
Plonna, 66
Plym, 67
Po, 131
Polota, 85
Porata, 115
Portva, 115
Practius, 167
Pravadi, 115
Pregel, 115
Primma, 154
Prims, 154
Pripet, 115
Pronia, 115
Prosna, 101
Prüm, 154
Pruth, 115
Purally, 115
Pydaras, 83
Pyramus, 154

Queiss, 158
Quenny, 145
Quipar, 177

Raab, 120
Rasa, 96
Rasay, 96
Ravee, 102
Raven, 102
Rea, 43

Rednitz, 95
Reen, 43
Rega, 43
Regen, 43
Regge, 43
Reno, 43
Reuss, 96
Rezat, 96
Rha, 43
Rhesus, 96
Rhine, 43
Rhion, 43
Rhodanus, 95
Rhodius, 95
Rhone, 95
Riaza; 96
Riga, 43
Riss, 96
Robe, 102
Rodach, 95
Rodau, 95
Rodden, 95
Roer, 168
Rohrbach, 168
Ross, 96
Rosslau, 96
Rötel, 96
Roth, 95
Rotha, 95
Rothaine, 95
Rother, 96
Rott, 95
Rottach, 95
Roubion, 102
Ruhr, 168
Rye, 43

Saale, 76
Saar, 55
Sabis, 59
Sabor, 59
Sabrina, 59
Saima (Lake), 119
Sal, 77
Salm, 166
Salo, 77

Salza, 151
Samara, 119
Sambre, 59, 119
San, 166
Saone, 119
Saraswati, 56
Saratovka, 56
Sarayu, 55
Sare, 55
Sark, 55
Sarnius, 56
Sarno, 56
Sarsonne, 56
Sarthe, 56
Sau, 59
Sauconna, 119
Save, 59
Savena, 59
Savezo, 59
Savio, 59
Savranka, 59
Sazawa, 98
Scaldis, 159
Scarr, 162
Scheer, 162
Scheldt, 159
Schie, 161
Schiltach, 159
Schmida, 53
Schnei, 52
Schondra, 99
Schozach, 99
Schunter, 99
Schupf, 101
Schussen, 99
Schutter, 99
Schwabach, 101
Schwale, 165
Schwalm, 166
Schwarza, 150
Schyrne, 162
Scius, 161
Scopas, 101
Seaton, 141
Seena, 166
Segre, 119

Segura, 119
Seille, 76
Seine, 119
Selle, 76
Selse, 151
Selune, 77
Sem, 119
Semoy, 119
Sempt, 119
Sena, 166
Senne, 166
Senus, 166
Seran, 56
Serchio, 55
Sered, 56
Sereth, 56
Serio, 55
Serre, 55
Serus, 55
Sessites, 98
Sestra, 99
Seugne, 119
Seva, 59
Sevan, 59
Severn, 59
Severus, 59
Sevre, 59
Sevron, 59
Shannon, 166
Sheaf, 101
Shere, 162
Shiel, 169
Shin, 166
Shira, 162
Sicoris, 119
Sid, 141
Sieg, 119
Sieve, 59
Sihl, 169
Silaro, 169
Sile, 169
Simmen, 119
Simmer, 119
Simois, 119, 169
Sinde, 23
Sitter, 141

Skerne, 162
Skippon, 101
Slaan, 77
Slaney, 77
Sneidbach, 52
Snyte, 52
Soar, 55
Soastus, 98
Soeste, 98
Soja, 119
Solman, 166
Somme, 119
Sora, 55
Sorg, 55
Sosna, 98
Sosterbach, 99
Sosva, 98
Souza, 98
Sow, 59
Söve, 59
Spean, 103
Spear, 103
Speier, 103
Spey, 103
Sprazah, 103
Spree, 103
Sprenzel, 104
Spressa, 104
Sprint, 103
Sprotta, 103
Stör, 58
Storas, 58
Stort, 58
Stour, 58
Streu, 58
Stroud, 58
Strumou, 171
Stry, 58
Stura, 58
Styr, 58
Suchona, 119
Suck, 59
Sucro, 59
Suevus, 101
Suippe, 101
Suire, 59

Sula, 165
Sulgas, 165
Sullane, 165
Sulm, 166
Sur, 55
Sura, 55
Sure, 55
Suren, 56
Suss, 98
Sutledge, 26, 98
Sutoodra, 98
Suusaa, 98
Suzon, 98
Svart, 150
Svir, 55
Swale, 165
Swelly, 165
Swilly, 165
Swords, 56
Syrmus, 171
Szala, 151

Ta, (Loch), 135
Tabuda, 135
Tacon, 107
Tamar, 135
Tamaris, 135
Tambre, 135
Tame, 135
Tamuda, 136
Tamyras, 136
Tana, 135
Tanagro, 136
Tanais, 135
Tanaro, 135
Tanger, 136
Tanus, 135
Taptee, 135
Tara, 149
Tardoire, 105
Tarf, 69
Tarisa, 149
Tarn, 149
Taro, 149
Tartaro, 105
Tartessus, 105

Tarth, 105
Tauber, 37
Tavda, 135
Tave, 135
Tavus, 135
Tavy, 134
Taw, 134, 135
Tay, 135
Teane, 135
Tearne, 149
Tearus, 179
Tees, 106
Teesta, 107
Teign, 135
Tema, 135
Teme, 136
Temes, 136
Tengs, 136
Termon, 155
Tescha, 107
Tessin, 107
Test, 107
Teviot, 135
Thames, 136
Thaya, 135
Theiss, 107
Thiele, 106
Thur, 37
Tiasa, 107
Ticino, 107
Till, 105
Tilse, 106
Tim, 135
Timao, 135
Timavus, 135
Tivy, 135
Tollen, 106
Tom, 135
Torre, 37
Tosa, 107
Töss, 107
Touse, 107
Touvre, 37
Towy, 36
Trachino, 71
Tragus, 70

Traun, 69
Trave, 69
Trebbia, 69
Treja, 70
Trent, 141
Trento, 141
Trome, 70, 155
Truentius, 141
Truim, 70, 155
Tura, 37
Turija, 37
Turuntus, 141
Twiste, 158
Tzna, 52

Uda, 176
Ufa, 176
Ui, 177
Uist, 158
Ulla, 89
Ullea, 89
Ullster, 89
Umbro, 28
Umea, 28
Unstrut, 58
Upa, 176
Ural, 40
Urius, 39
Urjumka, 122
Ursel, 78
Usk, 31
Uste, 158
Uxella, 31

Vaga, 63
Vagai, 63
Vahalis, 63
Vakh, 63
Varano, 78
Vardar, 79
Varde, 79
Vardre, 79
Varese (Lake), 78
Vartrey, 79
Vayah, 63

Vegiaur, 64
Vegre, 63
Vehne, 146
Veile, 90
Veistriz, 158
Vel, 90
Velez, 91
Velino, 91
Vellaur, 91
Vendée, 146
Vent, 145
Ver, 77
Verdon, 79
Vesdre, 158
Vesle, 158
Vever, 64
Veveyse, 64
Viaur, 63
Vie, 63
Vienne, 63
Vig, 63
Vilia, 90
Viliu, 90
Villa, 90
Vilna, 90
Vils, 91
Vindau, 146
Vipasa, 64
Vire, 77
Vistre, 158
Vistula, 158
Vlie, 65
Vliest, 66
Vliet, 66
Vodla, 34
Vosges, 63

Waag, 63
Waal, 63
Wandle, 146
Warnau, 77
Warta, 79
Watawa, 34
Waveney, 63
Waver, 63

Wear, 84
Weaver, 64
Wegierka, 64
Weichsel, 158
Welland, 90
Welse, 91
Wente, 179
Wern, 77
Werre, 77
Wers, 78
Wertach, 78
Wetter, 34
Wey, 63
Wick, 63
Wien, 63
Wigger, 63
Willy, 90
Windau, 146
Winderius, 146
Windermere (Lake), 146
Wipper, 64
Wislauf, 158
Wisloka, 158
Woder, 34
Worse, 78
Wölpe, 73
Wupper, 64
Wurdah, 79
Wyck, 177
Wye, 63

Xalon, 77
Xucar, 59

Yssel, 33
Ythan, 35

Zeyer, 59
Zorn, 56
Zna, 52
Zwettel, 158
Zwittau, 158
Zwittawa, 158